2nd Russia Japan USA Symposium on the Fundamental and Applied Problems of Terahertz Devices and Technologies

AA002779

(RJUS TeraTech 2013)

Journal of Physics: Conference Series Volume 486

Moscow, Russia
3 – 6 June 2013

ISBN: 978-1-63266-707-6
ISSN: 1742-6588

Printed from e-media with permission by:

Curran Associates, Inc.
57 Morehouse Lane
Red Hook, NY 12571

Some format issues inherent in the e-media version may also appear in this print version.

Copyright© by the Institute of Physics
All rights reserved.

Printed by Curran Associates, Inc. (2014)

For permission requests, please contact the Institute of Physics
at the address below.

Institute of Physics
Dirac House, Temple Back
Bristol BS1 6BE UK

Phone: 44 1 17 929 7481
Fax: 44 1 17 920 0979

techtracking@iop.org

Additional copies of this publication are available from:

Curran Associates, Inc.
57 Morehouse Lane
Red Hook, NY 12571 USA
Phone: 845-758-0400
Fax: 845-758-2634
Email: curran@proceedings.com
Web: www.proceedings.com

TABLE OF CONTENTS

Journal of Physics: Conference Series **486** (2014) 011001 doi:10.1088/1742-6596/486/1/011001

Preface

The 2nd Russia–Japan–USA Symposium "The Fundamental & Applied Problems of Terahertz Devices & Technologies" (RJUS TeraTech – 2013) Bauman Moscow State Technical University Moscow, Russia, 3rd–6th June, 2013

The 2nd Russia–Japan–USA Symposium "The Fundamental & Applied Problems of Terahertz Devices & Technologies" (RJUS TeraTech —- 2013) was held in Bauman Moscow State Technical University on 3rd–6th June 2013 and was devoted to modern problems of terahertz optical technologies. RJUS TeraTech 2013 was organized by Bauman Moscow State Technical University in cooperation with Tohoku University (Sendai, Japan) and University of Buffalo (The State University of New York, USA).

The Symposium was supported by Bauman Moscow State Technical University (Moscow, Russia) and Russian Foundation for Basic Research (grant # 13-08-06100-g).

RJUS TeraTech —- 2013 became a foundation for sharing and discussing modern and promising achievements in fundamental and applied problems of terahertz optical technologies, devices based on grapheme and grapheme strictures, condensed matter of different nature. Among participants of RJUS TeraTech — 2013, there were more than 100 researchers and students from different countries.

This volume contains proceedings of the 2nd Russia–Japan–USA Symposium "The Fundamental & Applied Problems of Terahertz Devices & Technologies".

Valeriy Karasik, Viktor Ryzhii and Stanislav Yurchenko

Bauman Moscow State Technical University

Content from this work may be used under the terms of the Creative Commons Attribution 3.0 licence. Any further distribution of this work must maintain attribution to the author(s) and the title of the work, journal citation and DOI.
Published under licence by IOP Publishing Ltd

Journal of Physics: Conference Series **486** (2014) 011001 doi:10.1088/1742-6596/486/1/011001

РФФИ
РОССИЙСКИЙ
ФОНД
ФУНДАМЕНТАЛЬНЫХ
ИССЛЕДОВАНИЙ

МГТУ
ИМ. Н.Э. БАУМАНА
1830

TOHOKU
UNIVERSITY

Rensselaer

University at Buffalo
The State University of New York

Journal of Physics: Conference Series **486** (2014) 011001 doi:10.1088/1742-6596/486/1/011001

Symposium chair
Anatoliy A Aleksandrov, Rector of BMSTU

Symposium co-chair
Valeriy E Karasik, Head of the Research and Educational Center "PHOTONICS AND INFRARED TECHNOLOGY" (Russia)

Invited Speakers

Taiichi Otsuji, Research Institute of Electrical Communication, Tohoku University, Sendai, Japan

Akira Satou, Research Institute of Electrical Communication, Tohoku University, Sendai, Japan

Michael Shur, Electrical, Computer and System Engineering and Physics, Applied Physics, and Astronomy, Rensselaer Polytechnic Institute, NY, USA

Natasha Kirova, University Paris-Sud, France

Andrei Sergeev, Department of Electrical Engineering, The University of Buffalo, The State University of New Your, Buffalo, NY, USA

Magnus Willander, Linkoping University (LIU), Department of Science and Technology, Linkopings, Sweden

Dmitry R Khohlov, Physical Faculty, Lomonosov Moscow State University, Russia

Vladimir L Vaks, Institute for Physics of Microstructures of Russian Academy of Sciences, Russia

Peer review statement

All papers published in this volume of *Journal of Physics: Conference Series* have been peer reviewed through processes administered by the proceedings Editors. Reviews were conducted by expert referees to the professional and scientific standards expected of a proceedings journal published by IOP Publishing.

Content from this work may be used under the terms of the Creative Commons Attribution 3.0 licence. Any further distribution of this work must maintain attribution to the author(s) and the title of the work, journal citation and DOI.
Published under licence by IOP Publishing Ltd

Boson peak in manganites

Boris P Gorshunov[1,2,3], Elena S Zhukova[1,2,3], Victor I Torgashev[4], Lenar S Kadyrov[1,2], Elizaveta A Motovilova[1,2], Florian Fischgrabe[5], Vasiliy Moshnyaga[5], Tao Zhang[6], Reinhard Kremer[7], Uwe Pracht[3], Sina Zapf[3] and Martin Dressel[3]

[1]A. M. Prokhorov General Physics Institute, Russian Academy of Sciences, Russia
[2]Moscow Institute of Physics and Technology (State University), Russia
[3]Physikalisches Institut, Universität Stuttgart, Germany
[4] Faculty of Physics, Southern Federal University, Russia
[5]Physikalisches Institut, Georg-August-Universität Göttingen, Germany
[6]Institute of Solid State Physics, Chinese Academy of Sciences, People's Republic of China
[7]Max-Planck-Institut fur Festkorperforschung, Germany

E-mail: gorshunov@ran.gpi.ru

Abstract. After the phase transition occurs in manganites $La_{1-x}Ca_xMnO_3$ ($x \geq 0.5$) into a dielectric phase with charge order, right below the corresponding temperature of $T_{CO} = 240\ K$ a series of absorption lines appear in the terahertz spectral range, only in case of commensurate x values. We show that the lines originate from acoustic phonons that become optically active by folding of the Brillouin zone. At even lower temperatures, a strongly asymmetric extra absorption band develops at frequencies around 20 cm^{-1} – 30 cm^{-1} corresponding to the position of the lowest-energy van Hove singularity in the reduced Brillouin zone. The band is assigned to the Boson Peak, i.e. to the excess of lattice vibrational states over the standard Debye contribution. The folded phonons and the boson peak do not show up for incommensurate calcium.

1. Introduction and background

Manganites of the family $R_{1-x}A_xMnO_3$ (R is a rare earth and A an alkaline element) provide a unique laboratory to study the physical properties of solids with several competing (or cooperating) order parameters. It is of general importance to understand the phase transitions to the various single- or multi-phase ground states [1] because eventually it can provide ideas for solving fundamental problems in the physics of electronically correlated systems. The aim of the present work was to analyze the mechanisms of phase transitions in $La_{1-x}Ca_xMnO_3$ by detailed exploration of the folded phonons dependent on the calcium content in the broad interval $0.5 \leq x \leq 1$. An important motivation was also to clarify the role of disorder and disorder-driven renormalization of the phonon density of states in the physics of overdoped manganites that are available mainly as polycrystals.

Content from this work may be used under the terms of the Creative Commons Attribution 3.0 licence. Any further distribution of this work must maintain attribution to the author(s) and the title of the work, journal citation and DOI.
Published under licence by IOP Publishing Ltd

Journal of Physics: Conference Series **486** (2014) 012001 doi:10.1088/1742-6596/486/1/012001

2. Results

Using a quasioptical BWO spectroscopic technique [2] in combination with Fourier-transform spectroscopy, we have measured terahertz-infrared response of a series of ceramics $La_{1-x}Ca_xMnO_3$ with $x(Ca) = 0.5; 0.6; 2/3 = 0.67; 0.7; 0.85; 0.9; 0.95; 0.98; 1$. We discover absorption bands with complex structures below TCO in the terahertz spectra (circle in figure1) for commensurate calcium concentration. We associate the structures with acoustical phonon modes which are optically activated in the charge ordered state where the crystal lattice forms a superstructure and the corresponding Brillouin zone is folded back. Most important, we observe a pronounced extra absorption feature with strongly asymmetric lineshape (arrow in figure 1), that we associate with the Boson Peak – a counterpart of the acoustic van Hove singularity, at the boundary of the folded Brillouin zone, that gains optical activity due to disorder. Our findings evidence that disorder effects must be taken into account when analyzing low-energy properties of overdoped manganites.

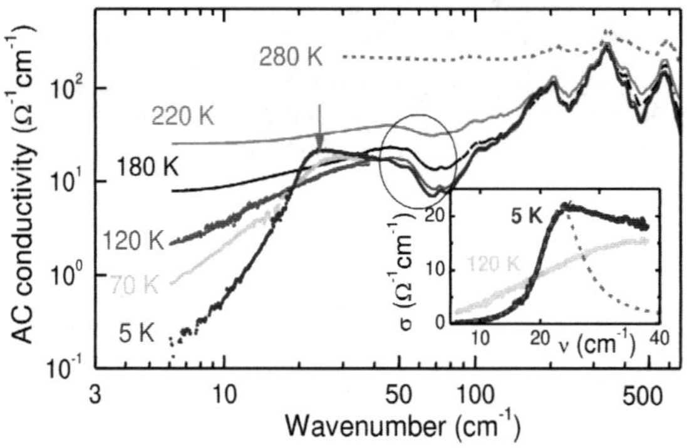

Figure 1. Temperature dependence of the terahertz-infrared conductivity of $La_{0.33}Ca_{0.67}MnO_3$ ceramic in the THz range. In the low-frequency part of the spectra the charge order transition at $T_{CO} = 240\ K$ is seen as dramatic change due to the opening of the energy gap. A circle indicates vibrational acoustic modes that become infrared active by Brillouin zone folding due to charge ordering. An arrow shows the Boson Peak. The inset demonstrates that it is not possible to fit the Peak lineshape by a Lorentzian curve (dashed line).

References

[1] Dagotto E 2003 *Nanoscale Phase Separation and Colossal Magnetoresistance* (Springer-Verlag, Berlin)

[2] Gorshunov B P et al. 2005 *Int. J. Infrared Millimeter Waves* **26** 1217

High precision spectroscopy and imaging in THz frequency range

Vladimir L Vaks

Institute for Physics of Microstructures of Russian Academy of Sciences, Russia

E-mail: vax@ipmras.ru

Abstract. Application of microwave methods for development of the THz frequency range has resulted in elaboration of high precision THz spectrometers based on nonstationary effects. The spectrometers characteristics (spectral resolution and sensitivity) meet the requirements for high precision analysis. The gas analyzers, based on the high precision spectrometers, have been successfully applied for analytical investigations of gas impurities in high pure substances. These investigations can be carried out both in absorption cell and in reactor. The devices can be used for ecological monitoring, detecting the components of chemical weapons and explosive in the atmosphere. The great field of THz investigations is the medicine application. Using the THz spectrometers developed one can detect markers for some diseases in exhaled air.

1. Introduction and background

The terahertz (THz) region is attractive for many well-known applications (biology and medicine, ecology and security). Until recently, it was difficult to efficiently generate and detect THz radiation. Most THz sources were either low-brightness emitters such as thermal sources, or cumbersome, single-frequency molecular vapor lasers.

Recently, however, there has been a significant progress in THz generation and detection due to achievements in nanotechnology and small-scale semiconductor industry.

2. Results

The present paper reports on microwave methods, devices and applications of THz spectrometry. We present the unique family of synthesizers operating in $667 \div 857\,GHz$, $789 \div 968\,GHz$, $882 \div 1100\,GHz$ [1]. Application of quantum superlattice devices has also advanced microwave spectroscopy methods to THz frequency range and yielded a family of THz spectrometers, based on non-stationary effects [2]. Medical and biological problems for THz technology involve, first of all, studying of biological compounds (DNA, proteins, sugars, *etc*) in solid phase and liquids. Dielectric properties of biomolecules in THz range are formed by low frequency vibrations, which specify collective motion of big atomic groups, and intermolecular hydrogen bonds, thus, providing information about molecular geometry and conformational flexibility and its chemical activity. The THz technique can also be used for studying of biological tissues based on dielectric function analysis. The main scientific purpose of this paper is development of a spectroscopic complex designed for studying of various biological systems (biomolecules, tissues, human organism). The complex includes spectrometers, based on free-damping polarization and fast passing effect and operating in near and far wave zones. The periodic switching the phase (or frequency) of radiation interacting in

Content from this work may be used under the terms of the Creative Commons Attribution 3.0 licence. Any further distribution of this work must maintain attribution to the author(s) and the title of the work, journal citation and DOI.
Published under licence by IOP Publishing Ltd

resonance with the medium leads to rising the processes of transient radiation and absorption, periodic appearing and decaying the macroscopic polarization induced [3]. The new frequency tunable sources of subTHz, THz and MIR radiation based on solid state harmonics generator and QCLs with frequency set not worse than 10^{-8} and power $10\ mcW - 10\ mW$, which meet requirements of high – precision spectroscopic measurements, are developed. The subTHz radiation source is developed on the solid state harmonics generator. For THz and MIR ranges the QCLs are used. They have a high output power and can generate radiation in pulse and continuous modes together with fast tuning of frequency. The room temperature PLL of THz QCL was developed and elaborated. Detection of the radiation in all three frequency ranges (subTHz, THz and MIR) are realized by the unique receiver block based on quantum semiconductor superlattices. The developed radiation sources are used for design and development of high-precision THz-imaging system and THz spectroscopy of biological objects.

One of the most important applications of the spectrometer elaborated is non-invasive medical diagnostics based on the analysis of the exhaled air [4]. We can identify some diseases with using the measurement of the characteristic gas-marker concentration in the exhaled air. Monitoring the variation of the NO concentration at the radiotherapy course was carried out. The absence of nitric oxide in exhaled air of healthy people and its appearance in exhaled air of oncopatients with cancer of lung which had radiotherapy are determined at clinical tests [4]. The experimental set up for precise spectral measurements of DNA in THz frequency range was developed and realized. The preliminary measurements of DNA spectra were carried out.

Acknowledgments
The work is supported by the grant of the **Government of the Russian Federation №** 11.G34.31.0066 **MedLab**, TeraDec project 047.018.005, NATO SfP – project 984068, Program of the Presidium of RAS “Fundamental sciences – to medicine”.

References
[1] Vaks V L, Panin A N, Koshurinov Yu I, Pavel'ev D G 2005 Development and design of a phase-locked loop in the subterahertz and terahertz ranges for a harmonic of the signal of a centimeter-wave synthesizer *Radiophysics and Quantum Electronics* **48** 831-836
[2] Vaks V L, Panin A N, Basov S A, Illyuk A V, Pripolzin S I, Paveliev D G, Koshurinov Yu I 2009 Nonstationary spectroscopy of the 1-2.5 THz frequency band with the use of solid-state devices *Radiophysics and Quantum Electronics* **52** 511-517
[3] Vaks V 2012 High-Precise Spectrometry of the Terahertz Frequency Range: The Methods, Approaches and Applications *J. Infrared, Millimeter and Terahertz Waves* **33** 43-53.
[4] Vaks V L, Domracheva E G, Sobakinskaya E A, Chernyaeva M B, Maslennikova A V 2012 Using the methods and facilities of nonsteady-state spectroscopy of the subterahertz and terahertz frequency ranges for noninvasive medical diagnosis *J. Opt. Tech.* **79**(2) 66–69

FDTD simulation of the electromagnetic field surface states in 2D photonic crystals

Kirill I Zaytsev[1], Vladimir S Gorelik[1,2], Alexey M Khorokhorov[1] and Stanislav O Yurchenko[1]

[1] Bauman Moscow State Technical University, Moscow, Russia
[2] Lebedev Physical Institute of the Russian Academy of Sciences, Moscow, Russia

E-mail: kirzay@gmail.com

Abstract. FDTD simulation was applied to study an optical field structure near the $2D$ photonic crystal surface. An effect of the optical field intensity redistribution was demonstrated for the first time. Spatial distribution of light intensity contains sharp peaks, focused in a certain parts of the crystal structure. The intensity peaks' magnitude could be very high. In case the incident light wavelength corresponds to the photonic crystal bandgap, the light could penetrate in a small depth into the crystal, but magnitude of the intensity peaks becomes several times higher in comparison to magnitude of the light intensity in the equal but homogeneous medium. Observed effect could be used for the enhancement of the light-matter interactions efficiency, for instance, for effective nonlinear conversion of light with photonic crystals, for Raman spectroscopy of small particles injected in crystals and etc.

1. Introduction and background

Photonic crystal is the structure characterized by the periodic spatial distribution of optical properties. The period of the optical properties' fluctuations is comparable to the wavelength of electromagnetic field [1],[2]. Usually the photonic crystals are formed of nanoparticles or nanowires. An example of natural photonic crystal is opal. It consists of the spherical quartz globules, organizing the face-centered-cubic lattice [3]. The diameter of the globules varies from 200 to 500 nm, and the approximate size if interglobular cavities is $10\ nm$. The properties of photonic crystal could be modified by filling the cavities with liquids or solids (figure 1).

The periodicity of the photonic crystal structure leads to the specificity in dispersion relation for photonic crystal. The photonic crystal is characterized with the presence of the bandgaps in the dispersion curves [4]-[7]. The easiest way to describe the principles of the bandgaps appearance is to use the 1D equivalent of the 3D globular face-centered-cubic lattice photonic crystal [8]. In the figure 2a the equivalent 1D periodic structure of the photonic crystal is presented. According to the Bloch's theorem, the solution of the Helmholtz equation in the periodic optical structure has the following form:

$$\tilde{u}(z) = \tilde{u}_K(z)\, exp(-iK(\omega)z)\,, \tag{1}$$

where $\tilde{u}_K(z) = \tilde{u}_K(z + \Lambda)$ is a periodic complex function with the period Λ, and $K(\omega)$ is the dispersion relation of the 1D photonic structure. It is easy to find rigorous solution for $K(\omega)$ [8]:

Content from this work may be used under the terms of the Creative Commons Attribution 3.0 licence. Any further distribution of this work must maintain attribution to the author(s) and the title of the work, journal citation and DOI.
Published under licence by IOP Publishing Ltd

(a) **(b)**

Figure 1. Structure of $3D$ opal globular photonic crystal visualized with the atomic force microscopy: (a) is the initial opal matrix and (b) is the opal matrix filled with the $NaNO_2$.

$$\cos(K(\omega)\Lambda) = \cos(k_1 a)\cos(k_2 b) - \frac{1}{2}\left(\frac{n_2}{n_1} + \frac{n_1}{n_2}\right)\sin(k_1 a)\sin(k_2 b), \qquad (2)$$

where a and b are the thicknesses of the layers, n_1 and n_2 are the refractive indexes of the media, k_1 and k_2 are the wavenumbers of the media. In the figure 2b the dispersion curve example is shown. The function $K(\omega)$ corresponds to the 1D photonic crystal structure with the following parameters: $a = 190\,nm$, $b = 60\,nm$, $n_1 = 4$, $n_2 = 3$. The dispersion relation contains several bandgaps, the region of the positive and negative group refractive indexes. The function $\tilde{u}_K(z)$ could be also found, for example, utilizing the perturbation theory.

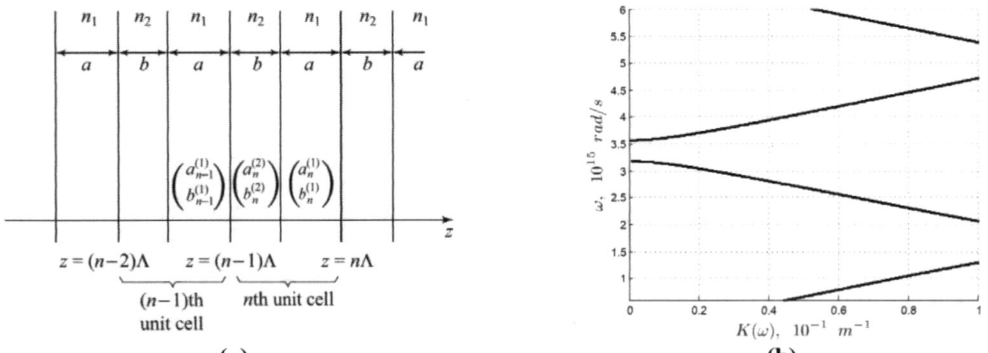

(a) **(b)**

Figure 2. Equivalent 1D photonic crystal characteristics: (a) is a scheme of equivalent $1D$ structure, (b) is dispersion curves of the $1D$ structure with $a = 190\,nm$, $b = 60\,nm$, $n_1 = 4$, $n_2 = 3$.

One-dimensional model of the photonic crystal (figure 2) could be used to describe the light distribution in the volume of medium. One of the most important problems of the photonic crystal technology is the examination of the effects of light interactions with photonic crystals, which appear near the air/crystal interface. The present paper contains the results of the numerical simulation of the electromagnetic field structure near the air/crystal boundary.

2. Results

The FDTD simulation [9] was implemented to study the spatial distribution of light intensity near the interface between the air and the $2D$ photonic crystal. $2D$ photonic crystal model is presented in the figure 3a. The medium consists of cylinders with the diameter $d = 250nm$. The refractive index of the cylinder medium is $n = 2$. The matrix of cylinders is filled with the medium, which refractive index is $n = 3$. The plane electromagnetic wave reached the surface of the crystal from the left side. The crystal size in all directions is much bigger than the incident light wavelength. It helps to prevent the appearance of the numerical simulation errors.

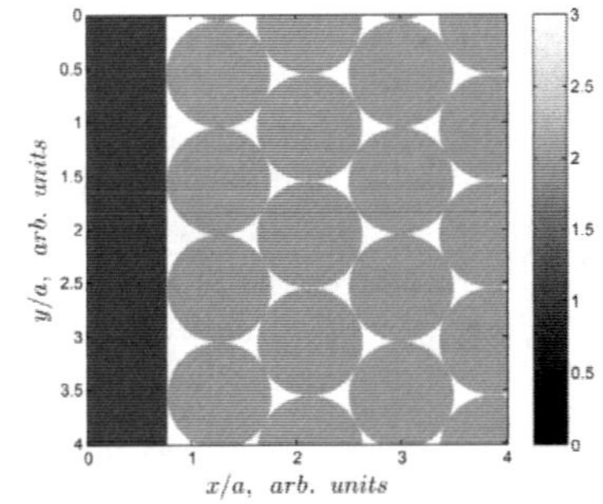

Figure 3. Spatial distribution of $2D$ photonic crystal refractive index.

After the FDTD simulation the spatial distribution of the electromagnetic field magnitude time dependence $u_{sim}(x, y, t)$ was detected. The function $u_{sim}(x, y, t)$ was used for calculation of the light intensity spatial distribution:

$$I_{sim}(x, y) = \frac{1}{T}\int_0^T |u_{sim}(x, y, t)|^2 dt, \quad T \gg \lambda_t, \tag{3}$$

where T is an integration time, λ_t is a period of the electromagnetic field oscillations in each point of the modelling space. Figure 4 contains the results of numerical modelling for several light wavelength ranging from 100 to 600 nm.

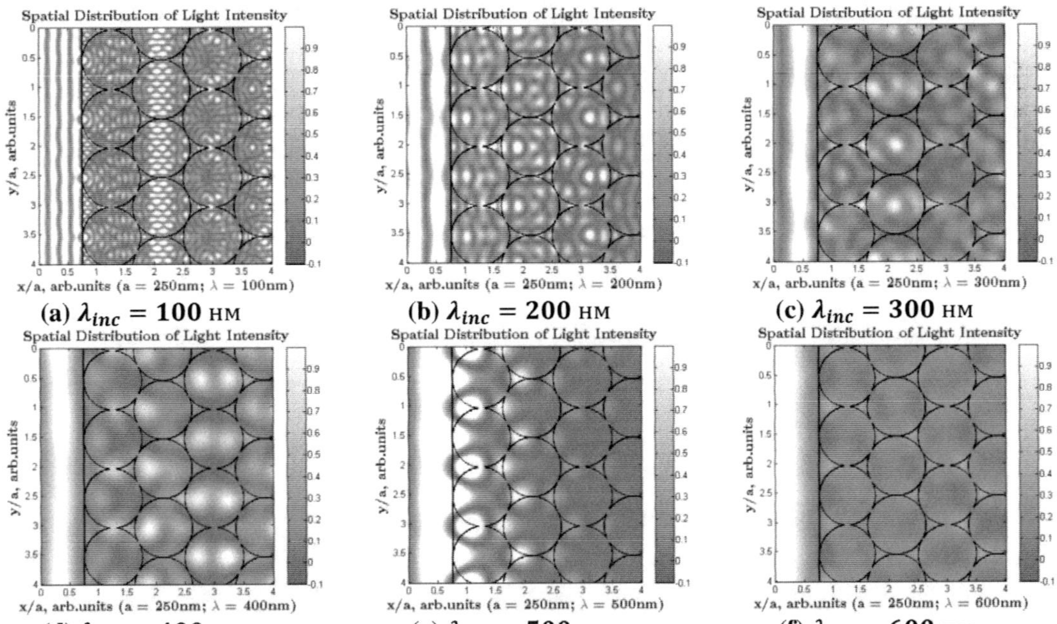

(a) $\lambda_{inc} = 100$ **HM** **(b)** $\lambda_{inc} = 200$ **HM** **(c)** $\lambda_{inc} = 300$ **HM**

(d) $\lambda_{inc} = 400$ **HM** **(e)** $\lambda_{inc} = 500$ **HM** **(f)** $\lambda_{inc} = 600$ **HM**

Figure 4. Spatial distribution of the electromagnetic field intensity near the interface between the air and the $2D$ photonic crystal.

In complete accordance with the Bloch's theorem (1) the function $I_{sim}(x,y)$ has a periodic character and decays with an increase of the crystal depth. All presented intensity distributions contain sharp peaks of light intensity. One could see, that the sharpest peaks and the highest decay of the field intensity corresponds to the wavelength $\lambda_{inc} = 500\ nm$ and it is equal to the wavelength of the crystal bandgap spectral position.

Conclusions

By means of the numerical FDTD simulation of the plane wave interaction with the air/2D-crystal interface, we have shown for the first time the effect of light intensity redistribution near the interface of photonic crystal. In case the incident light wavelength corresponds to bandgap of the photonic crystal, the peaks of the optical field intensity become several times higher than the intensity in equal homogeneous medium. The effect of the intensity redistribution is very important for a wide range of applications, in particular, for enhancement of nonlinear optical processes efficiency, for high efficient pumping of globular active structures, for increase of the Raman scattering on the particles injected in photonic crystals and others. It could become a basis for the creation of the new class of nonlinear optical media.

Acknowledgments

This work was partially supported by grants from the Russian Foundation for Basic Research (RFBR 14-02-00256; RFBR 13-02-90420; RFBR 13-02-00449; RFBR 14-02-00704).

References

[1] Yablonovitch E 1987 Inhibited Spontaneous Emission in Solid-State Physics and Electronics *Phys. Rev. Lett.* **58**(20) 2059 – 2062
[2] Sajeev J 1987 Strong Localization of Photons in Certain Disordered Dielectric Superlattices *Phys. Rev. Lett.* **58**(23) 2486 – 2489
[3] Gorelik V S 2008 Optics of globular photonic crystals *Laser Phys.* **18**(12) 1479 – 1500
[4] Ho K M, Chan C T, Soukoulis C M 1990 Existence of a Photonic Gap in Periodic Dielectric Structures *Phys. Rev. Lett.* **65**(25) 3152 – 3155
[5] Yablonovitch E, Gmitter T J, Leung K M 1991 Photonic band structure: The face-centered-cubic case employing nonspherical atoms *Phys. Rev. Lett.* **67**(17) 2295 – 2298
[6] Meade R D, Rappe A M, Brommer K D, Joannopoulos J D, Alerhand O L 1993 Accurate theoretical analysis of photonic band-gap materials *Phys. Rev. B* **48**(11) 8434 – 8437
[7] Hornreich R M, Shtrikman S, Sommers C 1994 Photonic band gaps in body-centered-cubic structures *Phys. Rev. B* **49**(16) 10914 – 10917
[8] Bunkin N F, Gorelik V S, Filatov V V 2010 Acoustic properties of globular photonic crystals based on synthetic opals *Phys. Wave Phenomena* **18**(2) 90 – 95
[9] Taflove A, Hagness S C *Computational Electrodynamics: The Finite-Difference Time-Domain Method. Second Edition* (Artech House, Boston, 2000) 852pp

RJUS TeraTech – 2013 IOP Publishing
Journal of Physics: Conference Series **486** (2014) 012004 doi:10.1088/1742-6596/486/1/012004

Hybrid mode-locked ultrashort-pulse erbium-doped fiber laser

Vladimir A Lazarev[1], Stanislav S Sazonkin[1], Alexey B Pniov[1], Konstantin P Tsapenko[1], Alexander A Krylov[2] and Elena D Obraztsova[3]

[1] Scientific and Research Centre "Photonics and IR-technics", Bauman Moscow State Technical University, 2nd Baumanskaya str., 5, Moscow 105005, Russia
[2] Fiber Optics Research Center of the Russian Academy of Sciences, Russia
[3] A. M. Prokhorov General Physics Institute of the Russian Academy of Sciences, Russia

E-mail: apniov@gmail.com

Abstract. One of the implementations of fs-laser with CNT-film for mode-locking is considered. Scheme of single-pulse, self-starting, stable mode-locked laser generation by appropriate polarization controllers adjustment is suggested. The mechanism of cavity length stabilization for a femtosecond fiber laser based on the pump source modulation is considered. Bandwidth of the feedback frequency stabilization system based on pump source modulation method is defined.

1. Introduction and background

In this paper, we employ the NPE as a fast saturation mechanism inside the ring cavity of the erbium-doped fiber laser mode-locked by carbon nanotubes (CNT). We compare the output parameters of ultra-short pulses in fiber laser based on a common ring scheme with a scheme that uses birefringence fiber. The NPE is initiated by including into the ring the birefringent "PANDA" – type silica-glass fiber that also possesses a polarizing effect – so-called PZ-fiber [3]. PZ-fiber also contributes to the suppression of the unwanted spectral spike at the wavelength of 1530 nm.

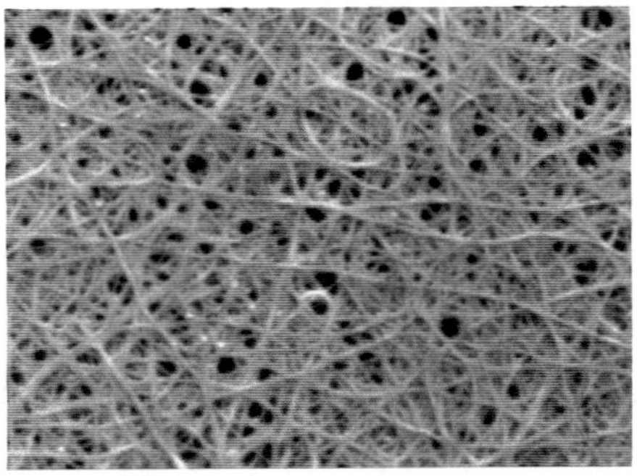

Figure 1. Carbon nanostructures.

Content from this work may be used under the terms of the Creative Commons Attribution 3.0 licence. Any further distribution of this work must maintain attribution to the author(s) and the title of the work, journal citation and DOI.
Published under licence by IOP Publishing Ltd

Journal of Physics: Conference Series **486** (2014) 012004 doi:10.1088/1742-6596/486/1/012004

The scheme of the mode-locked erbium-doped fiber ring laser is shown infigure 2. It consists of the active erbium-doped alumina-silicate fiber pumped by the 980 nm laser diode through a 980/1550 WDM. It should be noted, that erbium-doped fiber in the cavity possess positive group velocity dispersion (GVD) in the spectral band between 1500 and 1600 nm. Slow saturable absorber is based on carbon nanotubes incorporated in the polymer film (CNT-film) which is incorporated into the SAINT module.

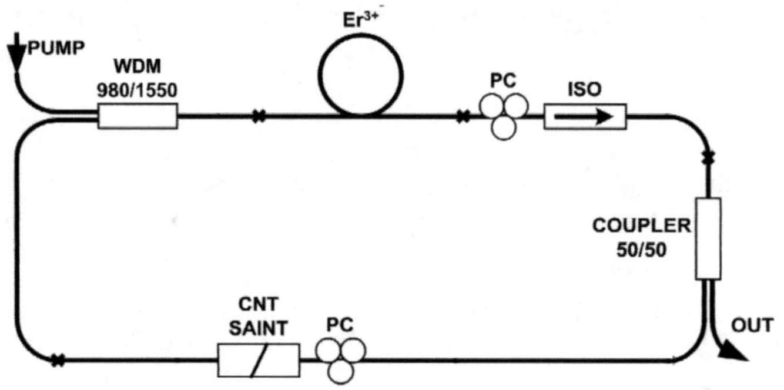

Figure 2. Scheme of the mode-locked erbium-doped fiber ring laser.

The setup of the mode-locked erbium-doped fiber ring laser with polarized birefringent fiber is quite similar the abovementioned standard ring scheme (figure 3). For effective realization of the NPE mechanism, the ring cavity contains two polarization controllers at both sides of the PZ–fiber (PZF). A commercial fiber isolator (ISO) ensures the unidirectional generation inside the cavity. The generated radiation is coupled out the cavity through the 3dB coupler.

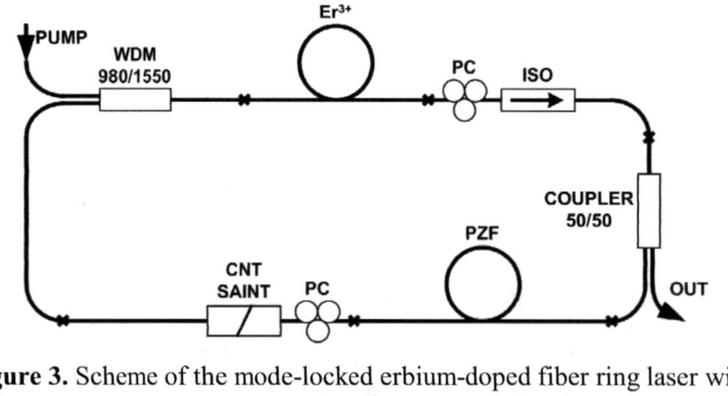

Figure 3. Scheme of the mode-locked erbium-doped fiber ring laser with PZ–fiber.

We have realized single-pulse, self-starting, stable mode-locked generation by appropriate polarization controllers adjustment. The ultrashort pulse characteristics have been measured through the varying of the CNT-film non-saturated transmission value (in the range 54–85%) and the pump power level. The shortest pulse duration was reached to be less than 250 fs in the case of the NPE mechanism implementation. Thus, the laser pulse-length was shortened by more than 2X in the presence of NPE mechanism inside the laser cavity.

Using of carbon nanostructures as a saturable absorber, can significantly extend the service life of the laser in comparison with similar lasers implemented on SESAM. Low power pulses does not have a strong influence on the life of the other components of the laser. On this basis one can conclude that hybrid mode-locked ultrashort-pulse erbium-doped fiber laser has a high service life. It's important to note that the price of carbon nanotubes is much lower than the cost of SESAM. It's possible to get the pulse generator with high power in combination with the Erbium-doped fiber amplifier. Operational life of the system will depend on the resources of the amplifier, which can be replaced by a reserve amplifier. The pulse duration of 250 fs allow to use this system in the optical frequency standards. Another significant advantage of the femtosecond laser is the ease assembly of the ring scheme.

Femtosecond lasers, emitting a sequence of ultrashort pulses, are a source of continuous equidistant narrow spectral lines, the interval between them is equal to the pulse repetition rate and the total length of the spectrum is determined by the duration of a single pulse. In other words, the femtosecond laser is not only a source of ultrashort pulses, but the comb generator of highly-monochromatic optical frequencies. If any component of the comb generator spectrum stabilize by the frequency standard, the absolute frequency values of all the components are known. In fact, it turns out the optical range, which can be used to effectively solve the problem of measuring optical frequencies.

2. Stabilized cavity length

One possible implementation of the frequency divider is a femtosecond laser with a stabilized cavity length. There is a variety of mechanisms for such stabilization: modulation of the pump source; using of the piezo-drive, which tensile and compressive optic fiber, as well as the use of electro-optic modulators. In this paper we study the mechanism of cavity length stabilization by means of pump source modulation. The mechanism is based on the length of the optical resonator, which is associated with a nonlinear refractive index of an optical fiber depended on the pump power. In this paper the band of feedback system of frequency stabilization was studied.

It is necessary to implement the feedback and adjust the length of the resonator fs-laser to stabilize the frequency of the fs-laser with He-Ne/CH_4 optical frequency standard (OFS). The mechanism is based on the length of the optical resonator, which is associated with a nonlinear refractive index of an optical fiber depended on the pump power:

$$l = nS, \qquad (1)$$

where n - the refraction index of an optical fiber [4], S - geometric path length. Presented expression shows that changing of the pump source power can change the optical length of the resonator. In an experiment we determined the bandwidth of the feedback for frequency stabilization system. The setup of this experiment is shown in figure 4.

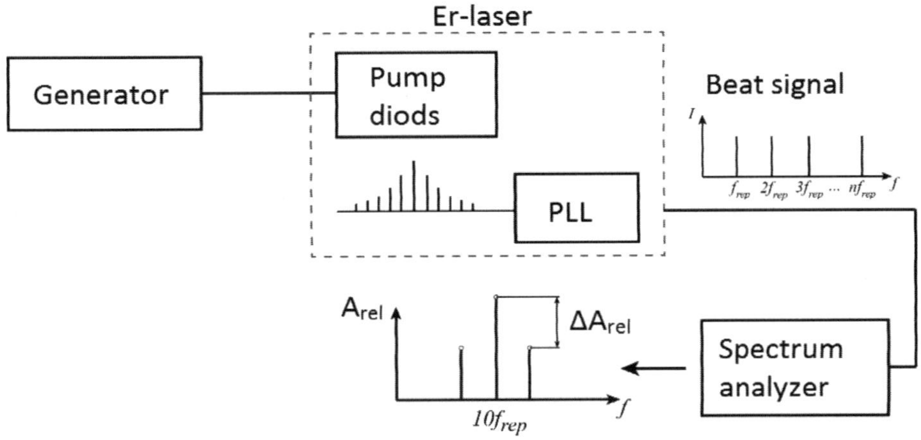

Figure 4. The experimental setup for determining the feedback bandwidth.

On this scheme sinusoidal signal generator is connected to a power supply pumping diodes fs laser. Beat signal optical frequency comb generated by a femtosecond erbium laser was recorded by the radiation detector, which is connected to an electrical spectrum analyzer. The pump power modulation result in the appearance of side components at each harmonic in the beat signal. The measurements were performed at the 30th harmonic of the beat signal. Difference between carrier amplitude and sideband amplitude was measured. Measurements results are plotted in figure 5.

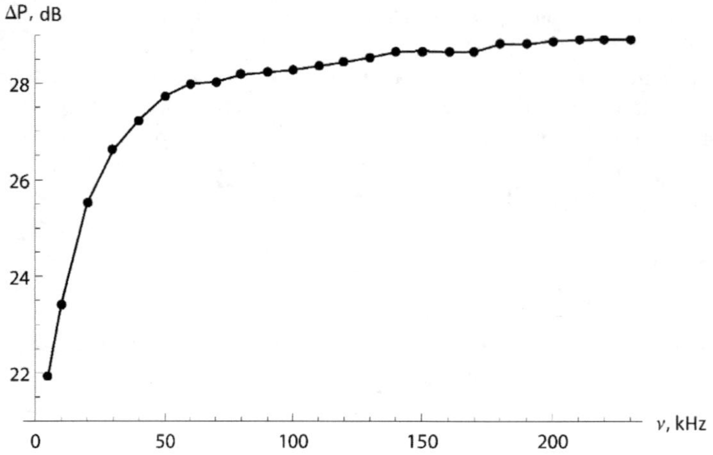

Figure 5. Results of measuring a bandwidth of the feedback for frequency stabilization system.

The graph shows that starting from 150 kHz, the changes are minor. Thus, the presented method allows to change the length of the cavity with feedback band of 150 kHz, which is better than the method of adjusting the cavity length based on the piezo-actuator, which feedback bandwidth is approximately 50 kHz.

3. Conclusion

In conclusion, we have compared pulse shortening by implementing the NPE mechanism inside the ring cavity of the CNT mode-locked erbium-doped fiber laser. We use an active erbium-doped fiber with positive group velocity dispersion (GVD) in the spectral band between 1500 and 1600 nm. The shortest pulse duration was less than 250 fs.

References

[1] Nelson L E, Jones D J, Tamura K et al. 1997 Ultrashort-pulse fiber ring lasers *Appl. Phys. B* **65** 277-294

[2] Set S Y, Yaguchi H, Tanaka Y, Jablonski M 2006 Ultrafast fiber pulsed lasers incorporating carbon nanotubes *IEEE J. Sel. Top. Quantum Electron.* **1** 137-146

[3] Koh J, Chen X, Li M J,et al. 2006 New polarizing optical fibers *Proc. SPIE* **6351** 63510L

[4] Akhmanov S A, Nikitin S Y 1998 *Physical Optics* (Oxford University Press. Oxford) p 656

Journal of Physics: Conference Series **486** (2014) 012005 doi:10.1088/1742-6596/486/1/012005

Terahertz rectification in a periodic two-dimensional electron plasma

Vyacheslav V Popov

Kotelnikov Institute of Radio Engineering and Electronics, Russia

E-mail: popov_slava@yahoo.co.uk

Abstract. The physics of terahertz rectification by periodic two-dimensional electron plasma is discussed. Ultrahigh responsivity of the terahertz detector with periodic two-dimensional electron plasma is predicted.

1. Introduction and background

Rectification of terahertz (THz) radiation due to plasmonic nonlinearities in the field-effect transistor (FET) with a two-dimensional electron channel, being predicted in [1], was demonstrated later in many papers (e.g., see [2][3]). Terahertz rectification by plasmons in the FET can be understood as a distributed resistive mixing effect [3] arisen due to applying THz voltage between the gate and source contacts. In order to apply THz voltage to the FET contacts, special antenna elements are needed [2][3]. In this paper, I discuss THz rectification in periodic two-dimensional electron plasma (2DEP) of a large area, which can be strongly coupled to THz radiation without using supplementary antenna elements. It is shown that, THz rectification by a periodic 2DEP results entirely from plasmonic nonlinearities with no THz voltage between the structure contacts involved. Ultrahigh responsivity of the plasmonic THz detection by a periodic 2DEP is predicted.

2. Results

Relatively dense 2DEP can be described by the hydrodynamic equations [4]

$$\frac{\partial V(x,t)}{\partial t} + V(x,t)\frac{\partial V(x,t)}{\partial x} + \frac{V(x,t)}{\tau} + \frac{e}{m}E(x,t) = 0, \tag{1}$$

$$e\frac{\partial}{\partial t}N(x,t) - \frac{\partial}{\partial x}j(x,t) = 0, \tag{2}$$

where $E(x,t)$ is the in-plane oscillating electric field (induced by the plasma oscillations in 2DEP) depending on the time t and coordinate x in 2DEP, $\tau = 1/\gamma$ is the electron scattering time, $j(x,t) = -eN(x,t)V(x,t)$ is the density of oscillating electric current, $N(x,t)$ and $V(x,t)$ are the hydrodynamic electron density and velocity in 2DEP, and e ($e > 0$) and m are the electron charge and effective mass, respectively. It is worth noting that equations (1) and (2) exhibit their nonlinear properties only due to distributed nature of plasma oscillations in 2DEP.

The plasmonic dc current in 2DEP, j_0, can be found by solving equations (1) and (2) using a perturbation approach with retaining the first and second powers of $E(x,t)$. For a travelling plasma wave with the electric field $E(x,t) = E_0\cos(\omega t - qx)$, where q is the wave vector, propagating along a homogeneous 2DEP with an equilibrium electron density $N_0(x) = N_0$,

Content from this work may be used under the terms of the Creative Commons Attribution 3.0 licence. Any further distribution of this work must maintain attribution to the author(s) and the title of the work, journal citation and DOI.
Published under licence by IOP Publishing Ltd

$$j_0 = -\frac{qe^3 N_0}{2m^2\omega(\omega^2+\gamma^2)} E_0^2, \tag{3}$$

which originates entirely from a nonlinear nature of equation (2) and can be interpreted as the plasmonic drag current $j_0 = j_0^{(drag)}$ The dc current induced by a travelling plasma wave in 2DEP with a periodic $N_0(x)$, $N_0(x) = N_0 + \Delta N_0 \cos(q_0 x)$, is determined by the same equation (3), however $j_0^{(drag)}$ vanishes for a standing plasma wave $E(x,t) = 2E_0 \cos(\omega t)\cos(qx + \varphi)$. In the latter case, dc current induced by plasma oscillations in 2DEP is

$$j_0 = -\frac{qe^3 \Delta N_0}{2m^2\gamma(\omega^2+\gamma^2)} E_0^2 \sin(2\varphi) \tag{4}$$

if $q_0 = 2q$ and $j_0 = 0$ if $q_0 \neq 2q$. The dc current described by equation (4) originates entirely from a nonlinear nature of equation (1) and can be interpreted as the plasmonic ratchet current $j_0 = j_0^{(ratch)}$ because $j_0^{(ratch)} = 0$ for $\varphi = 0$ ($j_0^{(ratch)}$ also vanishes in the case of a travelling plasma wave). For arbitrary electric field $E(x,t)$, both $j_0^{(drag)}$ and $j_0^{(ratch)}$ contribute to the total dc current excited in 2DEP. The plasmonic drag produces the major effect at sub-THz frequencies, while the plasmonic ratchet dominates at frequencies above $1\,THz$. Ultrahigh responsivity (up to $30\,kV/W$) of InGaAs/GaAs THz detector with a periodic 2DEP was confirmed by numerical simulations [5] and demonstrated experimentally [6] at room temperature.

References

[1] Dyakonov M, Shur M 1996 Detection, mixing, and frequency multiplication of terahertz radiation by two-dimensional electron fluid *IEEE Trans. Electron Devices* **43** 380-387

[2] Schuster F, Coquillat D, Videlier H et al. 2011 Broadband terahertz imaging with highly sensitive silicon CMOS detectors *Opt. Express* **19** 7827-7832

[3] Lisauskas A, Pfeiffer U, Öjefors E 2009 Rational design of high-responsivity detectors of terahertz radiation based on distributed self-mixing in silicon field-effect transistors *J. Appl. Phys.* **105** 114511-1 – 114511-7

[4] Popov V V 2011 Plasmon excitation and plasmonic detection of terahertz radiation in the grating-gate field-effect-transistor structures *J. Infrared Millim. Terahertz Waves* **32** 1178-1191

[5] Popov V V, Fateev D V, Otsuji T et al. 2011 Plasmonic terahertz detection by a double-grating-gate field-effect transistor structure with an asymmetric unit cell *Appl. Phys. Lett.* **99** 243504-1 – 243504-4

[6] Kurita Y, Ducournau G, Kobayashi K et al. 2013 Ultrahigh sensitive terahertz detection by asymmetric dual-grating gate HEMT structure *Int. Workshop Opt. Terahertz Sci. Technol. (OTST 2013), Kyoto, Japan, April 1-5, 2013: Extended Abstracts* 143

Terahertz frequency standard

Konstantin P Tsapenko[1,3], Vladimir A Lazarev[1,2], Stanislav O Leonov[1] and Aleksey B Pnev[1,2]

[1]Research and Educational Center "Photonics and Infrared Technology", Bauman Moscow State Technical University, 2nd Baumanskaya str., 5, Moscow, 105005, Russia

[2] Department of Laser and Optoelectronic Devices, Bauman Moscow State Technical University, 2nd Baumanskaya str., 5, Moscow 105005, Russia

E-mail: apniov@gmail.com

Abstract. An idea of highly stable terahertz pulse train designing by transforming stabilized optical femtosecond pulses using photoconducting antenna is offered. Experimental data obtained with THz pulsed spectrometer were used for creating a numerical mathematical model of THz comb.

1. Introduction

Along with progress in terahertz (THz) technology [1] the requirements of THz metrology are continuously increasing in various applications.

Frequency stability in THz region is relatively lower than in the optical and radio regions. Precise THz frequency measurements are required for spectrum analyzers, signal sources of standard frequency (THz clocks) and in wireless communication networks.

2. Link between optical and THz regions

Nowadays one of perspective ways of increasing frequency stability in the optical region is designing of Optical Frequency Standards. The accuracy of these standards is about 10^{-14} per second. This stability can be transferred to repetition frequency of femtosecond laser. Such sequence of optical pulses induces a stable THz pulse train in photoconductive (PC) antenna.

Er-fiber femtosecond laser was used for THz pulses generation. Repetition rate of optical pump pulses was 1 GHz and pulse duration – 100 fs. THz pulsed spectrometer allowed to determine the time-dependence of the single THz pulse amplitude. The sequence of THz pulses and its spectrum were modeled using the experimental waveform of single THz pulse. In the figure 1 induced THz pulse train is shown. In the figure 2 spectrum of single pulse is shown with red line, and THz-comb obtained from several pulses is shown with black lines.

Content from this work may be used under the terms of the Creative Commons Attribution 3.0 licence. Any further distribution of this work must maintain attribution to the author(s) and the title of the work, journal citation and DOI.
Published under licence by IOP Publishing Ltd

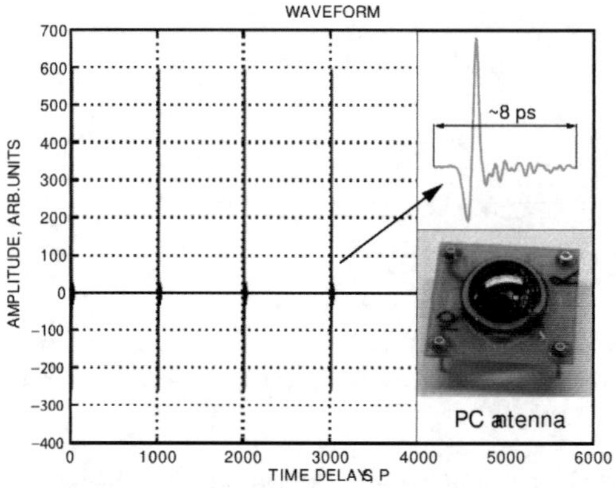

Figure 1. THz pulses induced in PC antenna.

The frequency comb is distorted by the noise of detector and changes in air composition along the THz beam path during the measurement process. The cause of particularly strong noise is water vapor fluctuations in the air. The spectral width of THz-comb was limited by the utilized method of THz pulse generation and by the duration of optical pulses.

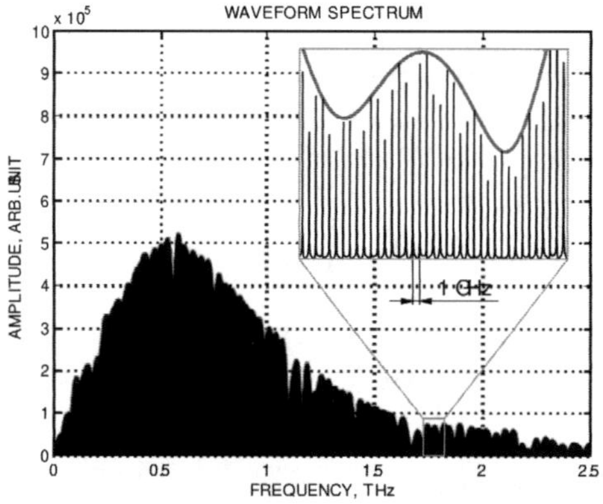

Figure 2. Spectrum of THz pulses.

One could use other methods of THz pulses generation: optical rectification [2], generation in air plasma [3]. It is possible to design multispectral frequency standard utilizing generation of THz pulses in air plasma. But the duration of optical pump pulses should be shorter than 10 fs. In this case the maximum frequency of THz spectrum can reach 30 – 40 THz.

CW-THz wave absolute measurement accuracy can reach 10^{-11} [4]. Also obtained comb can be used for THz-comb-referenced spectrum analyzers.

Future work will involve studying and stabilization of the THz frequency comb and designing of the stable frequency sources using THz frequency standard.

Acknowledgments
This work was partially supported by grants from the Ministry of Education and Science of Russian Federation (№ 16.513.11.3115).

References
[1] Lee Y-S 2009 *Principles of Terahertz Science and Technology* (Springer, New York, NY)

[2] Xu L, Zhang X-C and Heinz T F 1992 Terahertz beam generation by femtosecond optical pulses in electro-optic materials *Appl. Phys. Lett.* **61** 1784–86

[3] Hamster H, Sullivan A, Gordon S, White W and Falcone R W 1993 Subpicosecond, Electromagnetic Pulses from Intense Laser-Plasma Interaction *Phys. Rev. Lett.* **71** 2725–28

[4] Yasui T, Yokoyama S, Inaba H, Minoshima K, Nagatsuma T, Araki T 2011 Terahertz Frequency Metrology Based on Frequency Comb *IEEE J. Sel. Top. Quantum Electron.* **17** 191–201

Journal of Physics: Conference Series **486** (2014) 012007 doi:10.1088/1742-6596/486/1/012007

Challenges to create graphene terahertz lasers

Taiichi Otsuji[1], Akira Satou[1], Maxim Ryzhii[2], Vyachaslav Popov[3], Vladimir Mitin[4] and Victor Ryzhii[1]

[1] Research Institute of Electrical Communication, Tohoku University, Japan
[2] Computational Nano-Electronics Laboratoty, University of Aizu, Japan
[3] Kotelnikov Institute of Radio Engineering and Electronics (Saratov Branch), RAS, Russia
[4] Department of Electrical Engineering, University at Buffalo, SUNY, USA

E-mail: otsuji@riec.tohoku.ac.jp

Abstract. The gapless, linear energy spectrum of graphene can lead to population inversion and negative dynamic conductivity in the terahertz range when graphene is optically or electrically pumped, which makes it possible to create a new type of graphene terahertz lasers. This paper reviews recent advances in scientific challenges to create graphene-based terahertz lasers.

1. Introduction

Optical or injection pumping in graphene structures can lead to the interband population inversion and negative dynamic conductivity [1]-[4] (see figure 1), which provides an opportunity to create graphene terahertz (THz) lasers with pertinent laser cavity structures [5][6], including surface plasmon-polariton waveguides [7][8]. We discuss the importance of the carrier heating/cooling via optical phonons and of the excitation of surface plasmon polaritons to enhance the THz gain.

2. Results

Pumping graphene with rather high photon energy of ~eV extremely heats the carriers, hindering the population inversion [9]. Electrical injection with ~meV photons in graphene p-i-n junctions can minimize the carrier heating, and furthermore can give a possibility of carrier over-heating under high thermal conductivity conditions with high optical phonon decay rates (see figure 2), which substantially enhances the THz gain [10].

We theoretically discovered that when graphene carrier populations are inverted the excitation of graphene plasmons by the THz photons results in propagating surface plasmon polaritons with giant gain in a wide THz range (see figure 3) [7]. Furthermore, when graphene is patterned in a micro- or nano-ribbon array by grating gate metallization, the structure acts as an active plasmonic metamaterial, providing a super-radiant plasmonic lasing with giant gain at the plasmon modes in a wide THz frequency range [11].

Content from this work may be used under the terms of the Creative Commons Attribution 3.0 licence. Any further distribution of this work must maintain attribution to the author(s) and the title of the work, journal citation and DOI.
Published under licence by IOP Publishing Ltd

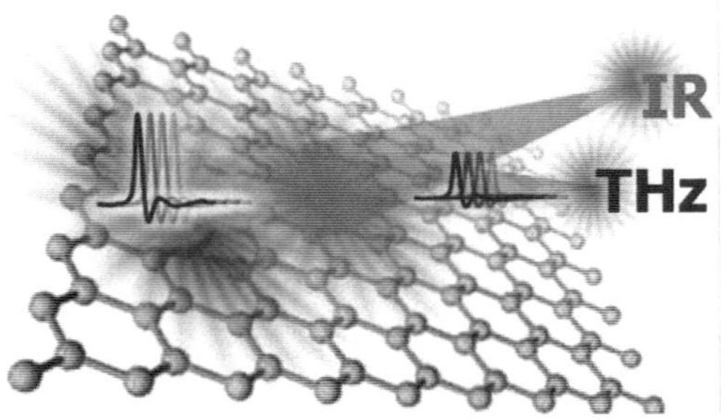

Figure 1. Terahertz gain in optically pumped grapheme This work is supported by JST-CREST, JSPS GA-SPR, JSPS Core-to-Core, Japan, NSF-PIRE-TeraNano, USA, and RFBRRAS, Russia.

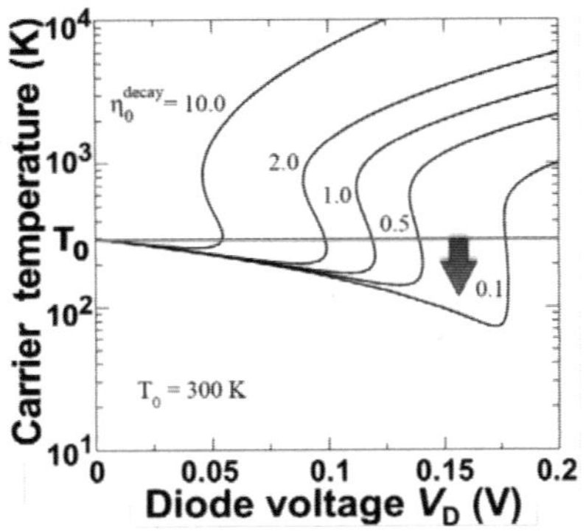

Figure 2. Effective carrier temperature vs. applied bias voltage for various optical phonon decay rate at 300K.

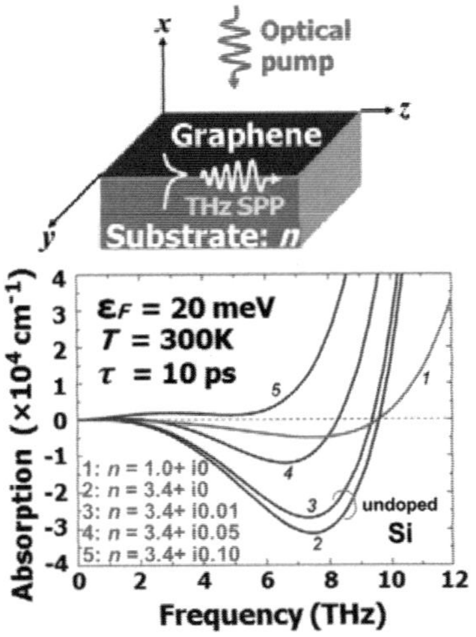

Figure 3. Excitation of surface plasmon polaritons in graphene and their absorption coefficient under population.

References

[1] Ryzhii V, Ryzhii M and Otsuji T 2007 *J. Appl. Phys.* **101** 083114
[2] Ryzhii M and Ryzhii V 2007 *Jpn. J. Appl. Phys.* **46** L151
[3] Boubanga-Tombet S et al. 2012 *Phys. Rev. B* **85** 035443
[4] Otsuji T et al. 2012 *J. Phys. D* **45** 303001
[5] Ryzhii V et al. 2009 *J. Appl. Phys.* **106** 084507
[6] Ryzhii V et al 2010 *J. Appl. Phys.* **107** 054505
[7] Dubinov A A et al. 2011 *J. Phys.:Cond. Mat.* **23** 145302
[8] Rana F 2008 *IEEE Trans. Nanotechnol.* **7** 91
[9] Satou A et al. 2013 *J. Appl. Phys.* **113** 143108
[10] Ryzhii V et al. 2011 *J. Appl. Phys.* **110** 094503
[11] Popov V V et al. 2012 *Phys. Rev. B* **86** 195437

Nondestructive testing of polymer composite materials using THz radiation

Egor V Yakovlev, Kirill I Zaytsev, Irina N Fokina, Valeriy E Karasik and Stanislav O Yurchenko

Research and Educational Center "Photonics and Infrared Technology"
Bauman Moscow State Technical University, 2nd Baumanskaya str., 5, Moscow, 105005, Russia

E-mail: yakov.egor@gmail.com

Abstract. This paper shows an ability to find internal defects in construction of polymer composite material using THz imaging. Using THz vision system resulted in an example of finding internal non-impregnated area in fiberglass.

1. Introduction and background

Terahertz radiation could have the unique ability to penetrate composites and identify defects such as voids, delaminations, mechanical damage and others [1]-[5]. The ability of finding heat damage in aircraft polymer composite materials, acquired during operation, by THz time-domain spectrometer (TDS) was studied in [1]. This work shows the ability of non-destructive control of technological defects, like non-impregnated area, in polymer composite materials, reinforced by glass fiber, by using THz continuous wave (CW) scanning vision system.

2. Results

Results of modeling operation of the CW vision system are shown in the figure 1. Interaction of THz radiation with construction of polymer composite material model was modeled by means of numerical solution of Maxwell's equations by finite-difference time-domain (FDTD) method.

Discretization parameters of space and time coordinates:

$$\Delta x = \Delta y = 0.003\ mm$$

$$\Delta t = \frac{\Delta x}{c\sqrt{2}} = \frac{\Delta y}{c\sqrt{2}} = 7.02\ fs$$

Content from this work may be used under the terms of the Creative Commons Attribution 3.0 licence. Any further distribution of this work must maintain attribution to the author(s) and the title of the work, journal citation and DOI.
Published under licence by IOP Publishing Ltd

Journal of Physics: Conference Series **486** (2014) 012008 doi:10.1088/1742-6596/486/1/012008

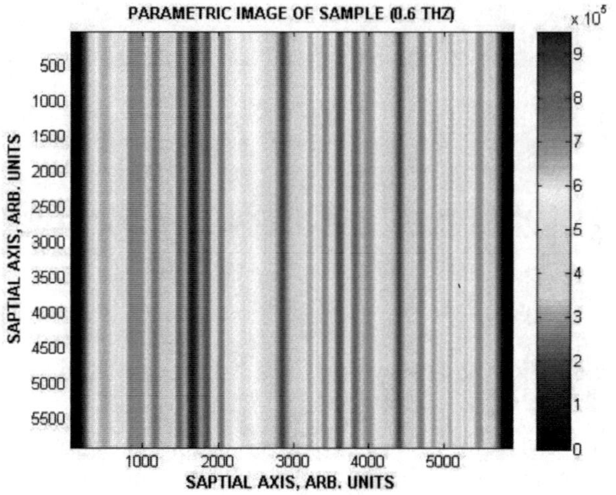

Figure 1. Modeling result of the THz vision system work, results are presented as form of parametric images.

Figure 1 presented the reflected from the sample model wave registered at an arbitrary distance from the surface of the object. Clearly visible irregularities reflected field in the areas of defects. Also it can be clearly seen that the deeper the defect area is placed the lower contrast of the image becomes. Construction model includes an ordered arrangement of fiberglass filaments, with a diameter 60 μm, placed in the polymer matrix. The model includes three simulated cylindrical defects, with a diameter 900 μm, as shown in the figure 2 and figure 3.

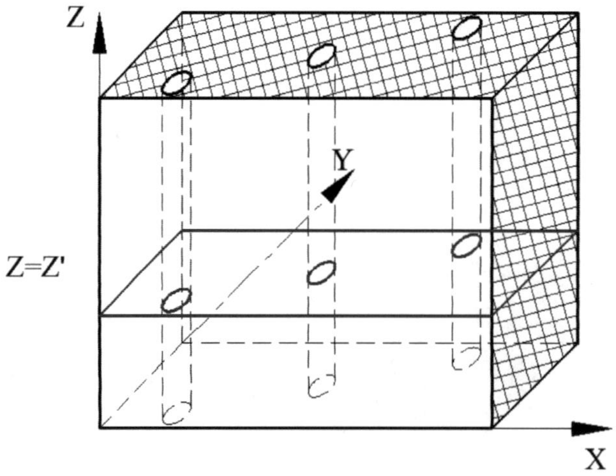

Figure 2. Product model of a polymer composite reinforced with fiberglass.

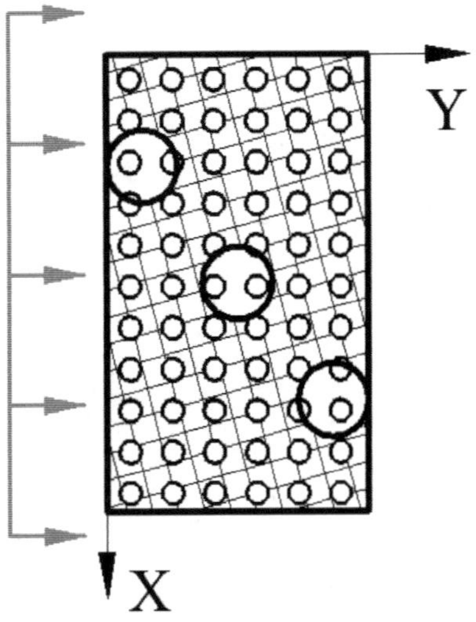

Figure 3. Cross-section Z=Z' of the model.

To verify experimentally the result of modeling, a sample of polymer composite material glass fiber reinforced was studied.

Several defects were simulated in the sample (figure 4). Defects are linear non-impregnated area, with the dimensions: length 15 mm width 3 mm depth 2 mm. THz radiation wavelength used for the control of sample quality was 1 THz.

The figure 5 showed result of scanning a sample of this polymer composite materials. The experiment was performed on the scanning THz CW vision system.

Figure 4. The sample of polymer composite material reinforced by fiberglass, with simulated defects. (scan area is highlighted in green).

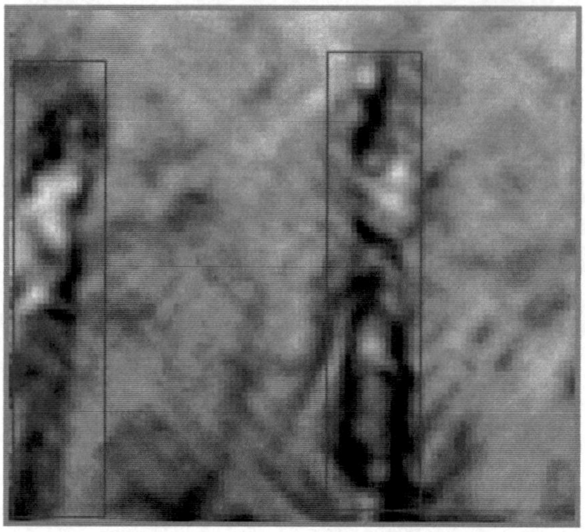

Figure 5. The result of scanning a sample of polymer composite reinforced by fiberglass using THz vision system,presented in the form of images (the defect area – non-impregnated stripes – is highlighted in red).

A potential of THz CW imaging system to detect the defects in polymer constructions was studied numerically and experimentally. Obtained images showed that such defects as non-impregnated area can be imaged clearly.

Acknowledgement

This work was partially supported by a grant from the Ministry of Education and Science of Russian Federation (№ 14.B37.21.0898, № 14.B37.21.1282) and from the Russian Foundation for Basic Research (RFBR 12-08-31104; RFBR 12-08-33112).

References

[1] Stoik C, Bohn M, Blackshire J 2008 Nondestructive evaluation of aircraft composites using transmissive terahertz time domain spectroscopy *Opt. Express* **21** 17039

[2] Wietzke S, Jordens C, Krumbholz N, Baudrit B, Bastian M, Koch M 2007 Terahertz imaging: a new non-destructive technique for the quality control of plastic weld joints *J. Euro. Opt. Soc.* **2** 07013

[3] Redo-Sanchez A, Karpowicz N, Xu J, Zhang X 2006 Damage and defect inspection with terahertz waves *The Fourth International Workshop on Ultrasonic and Advanced Methods for Nondestructive Testing and Material Characterization (UMass Dartmouth, N. Dartmouth, MA)*

[4] Rutz F, Hasek T, Koch M, Richer H, Ewert U 2006 Terahertz birefringence of liquid crystal polymers *Appl. Phys. Lett.* **89** 221911

[5] Reiten M, Hess L, Cheville R 2006 Nondestructive evaluation of ceramic materials using terahertz impulse ranging *Proc. SPIE* 617905 1-8

Electronic ferroelectricity in carbon-based systems: from reality of organic conductors to promises of polymers and graphene nano-ribbons

Natasha Kirova[1] and Serguei Brazovskii[2]

[1] LPS, CNRS-UMR 8502, Bat. 510,Université Paris-sud, 91405 Orsay cedex, France
[2] LPTMS, CNRS-UMR 8626, Bat. 100,Université Paris-sud, 91405 Orsay cedex, France

E-mail: nathalie.kirova@gmail.com

Abstract. Ferroelectricity is a rising demand in fundamental and applied solid state physics. Ferroelectrics are used in microelectronics as active gate materials, in capacitors, electro-optical-acoustic modulators, etc. There is a particular demand for plastic ferroelectrics, e.g. as a sensor for acoustic imaging in medicine and beyond, in shapeable capacitors, etc. Microscopic mechanisms of ferroelectric polarization in traditional materials are typically ionic. In this talk we discuss the electronic ferroelectrics - carbon-based materials: organic crystals, conducting polymers and graphene nano-ribbons. The motion of walls, separating domains with opposite electric polarisation, can be influenced and manipulated by terahertz and infra-red range optics.

1. Introduction

Phenomenon of the ferroelectricity referrers to an ability of certain materials to develop a spontaneous electric polarization which may have several allowed directions, hence it can be altered by an external electric field. Ferroeletrics have immense range of applications attracting related fundamental studies. The ferroelectric materials are the most universal: beyond their own virtues, they serve as pyroelectrics if treated by poled quenching, or as piesoelectrics if disordered.

The known microscopic mechanisms for the ferroelectric polarization are typically ionic: ions displacements from a centro-symmetric position in displacive ferroelectrics like most known PZT - $Pb(Zr_xTi_{1-x})O_3$, ordering of ions among equivalent potential minima (order-disorder type of the feroelectricity like in $BaTiO_3$), neutral-ionic transition where the charge disproportionation takes place among atoms in non-symmetric positions. There is a particular demand for plastic ferroelectrics, e.g. as a sensor for acoustic imaging in medicine and beyond, in shapeable capacitors, etc. Until now, this demand is satisfied by exploiting composite materials based on powders of ferroelectric oxides in a polymeric matrix. A single example of a purely polymeric FE is the saturated (that means not being active either electronically or optically) polymer Poly(vinylidene flouride) (PVDF). This material shows the ferroelectricity of a conformational origin, i.e. via chain rotation. Polarizabilities of existing plastic ferroelectrics are rather low, providing the dielectric permittivity $\varepsilon \sim 10$, hence resulting in a modest efficiency of devices. A natural attempt is to consider synthetic conductors, like organic crystals or conducting polymers, which π- electronic systems can greatly enhance the dielectric permittivity. In this case, the major polarization comes from redistribution of electronic density,

Content from this work may be used under the terms of the Creative Commons Attribution 3.0 licence. Any further distribution of this work must maintain attribution to the author(s) and the title of the work, journal citation and DOI.
Published under licence by IOP Publishing Ltd

Journal of Physics: Conference Series **486** (2014) 012009 doi:10.1088/1742-6596/486/1/012009

resulting in amplification of dielectric permittivity ε by a factor of $(\omega_p/\Delta)^2 \sim 10^2$ (ω_p is the plasma frequency, Δ is the gap in the electronic spectrum).

The contemporary discovery of the ferroelectricity in organic conductors [1] has brought to life a new mechanism of the predominantly electronic origin, which opens new scales of the magnitude and the rapidity of the effect. The high dielectric susceptibility, $\varepsilon \sim 10^3$, coexists with an unusually high conductivity, which gives rise to a "ferroelectric narrow-gap semiconductor". The intrinsic conductivity eliminates the usual hysteresis, which may lead to fast repolarization, which can be tuned by terahertz and infra-red optics.

2. Ferroelectricity in π-conjugated systems

The picture of the ferroelectricity implies existence of a doubly degenerate ground state in the system. For quasi one dimensional conductors, the microscopic picture of the ferroelectricity is based on two coexisting symmetry lowering effects: the dimerization of bonds and the dimerization of sites, see figure 1. The important requirement is that one of these two effects should be spontaneous.

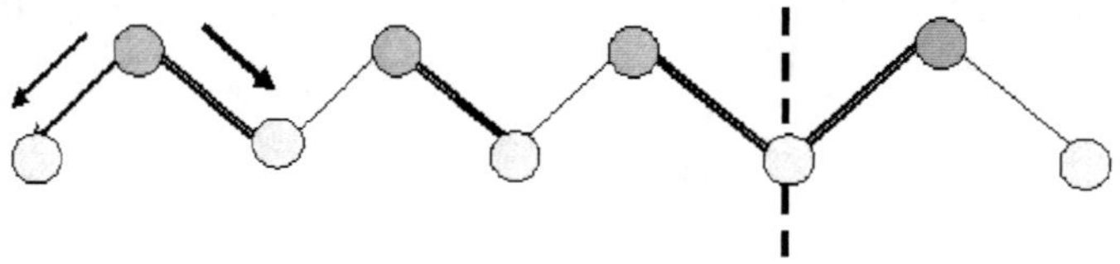

Figure 1. Joint effect of site and bond dimerizations for the ferroelectric ground state. The arrows show dielectric moments of bonds: they do exist because of the site inequivalence and they are different in magnitudes because of inequivalence of bonds. Vertical line indicates the position of the soliton – ferroelectric domain wall.

The combination of the two effects removes the inversion and the mirror symmetry, eliminating the glide plane. Nevertheless, the doubly degenerate ground state is preserved, which immediately gives rise to existence of solitons [1]-[3]. We may consider solitons as ferroelectric domain walls separating domains with opposite electric polarization. A special experimental advantage follows that the ac electric field alternates polarization by moving charged solitons. Through solitons' spectral features in terahertz and infra-red ranges, it opens a special tool of electro-optical interference. In return, the physics of these exotic solitons will serve to describe transient processes in ferroelectric polymers [3].

In perspectives of applications, crystalline organic materials may be too fragile, also their transition temperatures to the ferroelectric state are still in the cryogenic range. So it is tempting to find a similar effect in more robust and technological materials like conjugated polymers.

Major activity in physics and applications of conjugated polymers has been centering on their ability to emit light in the optical range of about 2eV (see e.g. [4] and references therein). These optically active materials are already close to applications and even commercialized. If one succeeds using their fast π-electrons to perform also the ferroelectricity, it will be possible to use the transient ferroelectric processes to monitor the visible range optics and vice versa.

A secure guess is to look for the so called $(AB)_x$ polymer – see figure1, proposed theoretically in early 80s [2]. Indeed, the first member of this potentially vast family of "di-substituted polyacetylenes" has already been synthesized, studied for nonlinear optical properties, but not tested yet for the low frequency response which could have recovered the ferroelectricity.

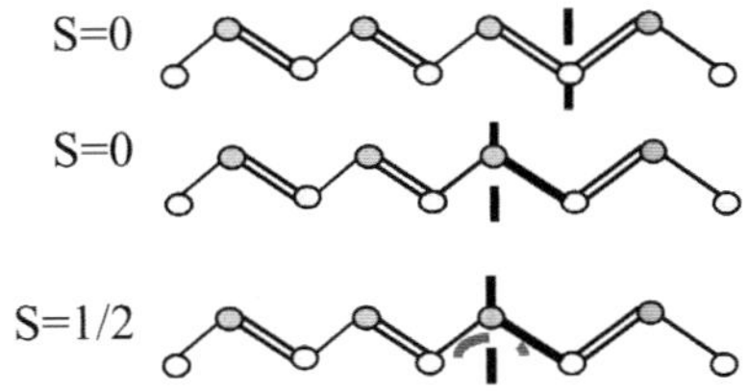

Figure 2. Three types of topological solitons in $(AB)_x$ polymeric chain.

The double degeneracy of the ground state is preserved, giving rise to three different topological solitons - bond dimerization kinks, see figure 2. But unlike the standard polyacetylene chain, all the three types of solitons carry variable, non integer charges [2],[3]. So, for this combined Peierls state, there is no total spin-charge separation, contrary to traditional 1D Peierls and Hubbard models. Thus the earlier neutral spin-carrying soliton, S=1/2, acquires the charge $Q_s=e(2/\pi)\tan^{-1}(\Delta_{ex}/\Delta)$. The spinless solitons (S=0) change their charges according to $Q_{\pm}=Q_s\pm e$.

The only known existing di-substituted polyacetylene has been synthesized a decade ago. Later on it attracted an attention in view of very promising applications as a light emitting material, even the lasing effect has been demonstrated [5]. The detailed optical characterization [6] provides an indirect proof for presence of the spontaneous bonds dimerization via spectral signatures of solitons. An "accidental" origin of the success to get the Peierls effect comes from a truly weak difference of alternating radicals: they are the phenyl rings which differ only by a side group attached most distantly from the carbon backbone bearing the π-electrons. Then the build-in site dimerization gap is small and the system is provoked to add the bond dimerization contribution to the gap. There was no check for the ferroelectricity yet: it seems plausible to be tried and realistic to be discovered.

We consider also a possibility of ferroelectricity for zigzag edges of graphene ribbons, see Figure 3. The build-in dimerization in this case is originated by different chemical surrounding of edge and bulk carbon atoms. The spontaneous dimerization can be the result of electron-phonon interactions and Peierls transition along the edges.

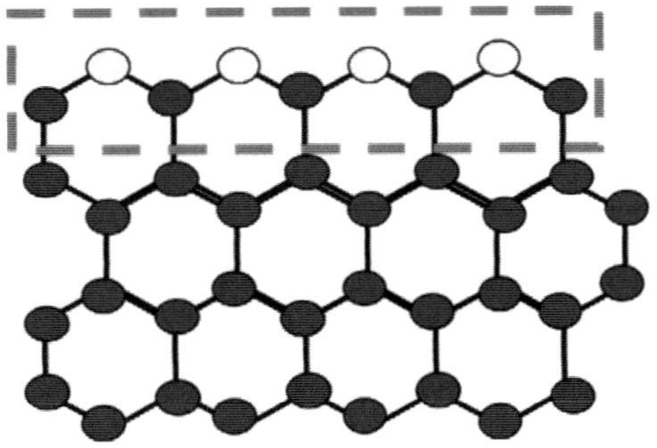

Figure 3. Possible ferroelectricity at the edges of graphene nano-ribbons.

3. Conclusions

π-conjugated systems can support the electronic ferroelectricity. The design is symmetrically defined and can be previewed.

Conductivity and/or terahertz activity of electronic ferroelectrics will add more functionality to their ferroelectric states. Re-polarization ferroelectric walls being influenced by terahertz and infra-red

Journal of Physics: Conference Series **486** (2014) 012009 doi:10.1088/1742-6596/486/1/012009

range optics and their studies will add to reconciliation of applied and fundamental physics of carbon based systems.

References

[1] Monceau P, Nad F and Brazovskii S 2001 *Phys. Rev. Lett.* **86** 4080
[2] Brazovskii S, Kirova N 1981 *JETP Lett.* **33** 4
[3] Kirova N, Brazovskii *Physica B*
[4] Kirova N 2008 *Polymer Int.* **57** 678
[5] Gontia I, Frolov S V, Liess M, Ehrenfreund E, Vardeny Z V, Tada K, Kajii H, Hidayat R, Fujii A, Yoshino K, Teraguchi M and Masuda T 1999 *Phys. Rev. Lett.* **82** 4058
[6] Luer L, Manzoni C, Cerullo G, Lanzani G, Vardeny Z V 2007 *Chem. Phys. Lett.* **444** 61

Novel technique for medium permittivity profile reconstruction using THz pulsed spectroscopy

Kirill I Zaytsev, Nikita V Chernomyrdin and Valentin I Alekhnovich

Bauman Moscow State Technical University, 2nd Baumanskaya str., 5, Moscow, 105005, Russia

E-mail: kirzay@gmail.com

Abstract. A new method for permittivity profile reconstruction based on terahertz time-domain spectroscopy signal processing is presented. Reconstruction is accomplished in two steps. First, the sample pulse function is reconstructed using sample time-domain reflection data. Low and high frequency noise filtering and the interpolation of the pulse function at low frequencies are then applied. Second, an invariant embedding technique is used to calculate the dielectric permittivity profile based on the sample pulse function. Method was tested experimentally on samples with known permittivity profiles. The algorithm is stable to additive Gaussian white noise as shown using mathematical modeling based on the finite-difference time-domain technique (FDTD). Possible applications of this permittivity profile reconstruction technique are discussed.

1. Introduction and background

Terahertz (THz) radiation is located between the infrared and the microwave regions of the electromagnetic spectrum, from about 0.1 to 10.0 THz. THz waves have some specific properties, such as high penetration depth into dielectric materials, and the ability to interact with both vibrational and rotational molecular levels [1]. THz radiation is non-ionizing in nature. Because of these properties, there are many potential applications of THz technology [2]-[5].

THz time-domain spectroscopy (TDS) utilizes short pulses of THz radiation with a wideband spectrum (from 0.1 to 3.0 THz) to measure the THz material parameters of various media. The electric field of a THz pulse is detected with a very high time resolution (about 50.0 fs) after the transmission of the pulse through the sample or the reflection of the pulse from the surface of the sample. By applying a fast Fourier transform to the detected time-domain data, it is possible to analyze the complex transmission or reflection coefficients of the sample in a wide frequency range.

Time-domain data of TDS system can also be used to reconstruct the permittivity profile in a sample, i.e., the dependence of the sample dielectric permittivity on depth. This reconstruction method aids in the study of the internal structure of a sample, known as THz tomography or T-Ray tomography [6]. THz tomography is expected to be useful in medical diagnosis, nondestructive evaluation of construction materials, inspection of art objects, and other applications. This paper describes a new way to solve this inverse problem. The results of implementing the new algorithm and its stability with respect to noise are shown.

Several assumptions concerning media properties are made in the present work. The medium of interest is assumed to be non-dispersive and non-absorbing. Sample optical properties are assumed to be constant in the lateral direction and vary only with depth. The media of interest must possess no

Content from this work may be used under the terms of the Creative Commons Attribution 3.0 licence. Any further distribution of this work must maintain attribution to the author(s) and the title of the work, journal citation and DOI.
Published under licence by IOP Publishing Ltd

Journal of Physics: Conference Series **486** (2014) 012010 doi:10.1088/1742-6596/486/1/012010

optical non-linearity. The medium is assumed to be thick such that multiple reflections in sample layers do not impact the reflected signal data (i.e. the number of pulses in the sample response). Notice, while studying materials with rather high dispersion of optical properties we are reconstructing sample permittivity profile, which corresponds to the average of the sample permittivity in the entire bandwidth of the THz pulse spectrum.

We consider all stages of permittivity profile reconstruction using this algorithm. We discuss all methods for solving the problems of invariant embedding technique generalization to THz spectroscopy.

2. Results

2.1. Algorithm description

Sample dielectric permittivity profile reconstruction is accomplished in two steps: reconstruction of the sample pulse function $R(t)$ based on TDS system signals and the reconstruction of the sample permittivity profile $\varepsilon(z)$ based on a sample pulse function $R(t)$. Two signals need to be obtained for sample permittivity profile reconstruction: $E_s(t)$ is the signal reflected from the sample of interest, and $E_r(t)$ is the reference signal reflected from the reference surface with a very high and homogeneous reflectivity in a wide spectral range. A planar gold mirror was used to obtain the reference signal.

The sample pulse function $R(t)$ is the medium response to the influence of an incident electromagnetic wave with $\delta(t)$ amplitude time dependence, where $\delta(t)$ is a Dirac delta-function. The sample pulse function is also the result of an inverse Fourier transform applied to the sample amplitude reflectivity $\tilde{R}(\nu_t)$.

A technique utilizing a Wiener filter $\tilde{F}(\nu_t)$ was applied in previous research [7] to process TDS system data. The formulation of a derived Wiener filter with several modifications is used to solve the deconvolution problem in Fourier-domain and to obtain sample spectral reflectivity $\tilde{R}(\nu_t)$:

$$\tilde{R}(\nu_t) = \frac{\tilde{E}_s(\nu_t)}{\tilde{E}_r(\nu_t)} \frac{\left(\frac{\tilde{E}_r(\nu_t)}{\max\limits_{\nu_t}\{\tilde{E}_r(\nu_t)\}}\right)^2}{\left(\frac{\tilde{E}_r(\nu_t)}{\max\limits_{\nu_t}\{\tilde{E}_r(\nu_t)\}}\right)^2 + \frac{N(\nu_t)}{S(\nu_t)}}, \quad (1)$$

where $N(\nu_t)$ is a noise power spectrum model, and $S(\nu_t)$ is a signal power spectrum model. White Gaussian noise power spectrum as a basis for the function $N(\nu_t)$. The function $S(\nu_t)$ could be easily defined with Gaussian monopulse model of TDS signal. The equation (1) is negligible in a frequency range where the noise power spectrum model $N(\nu_t)$ exceeds the signal power spectrum model $S(\nu_t)$. Conversely, if $S(\nu_t)$ exceeds $N(\nu_t)$, the function $\tilde{F}(\nu_t)$ is close to unity. Smooth filter edges are also provided. This filtering procedure allows us to obtain as much data on the sample reflectivity as possible, and contains smoothed filter edges to prohibit the emergence of the Gibbs noise in the sample pulse function [8],[9]. Obtained sample reflectivity depends only on the dielectric permittivity $\varepsilon(z)$ and the effective conductivity $\sigma(z)$ of the sample, and does not depend on the properties of TDS system signals.

The second problem in the pulse function reconstruction procedure is a reflectivity interpolation at low frequencies. To reconstruct a sample permittivity profile with the invariant embedding technique, it is essential to know the sample pulse function spectral components at low frequencies, excluding the zero point of the discrete Fourier-domain. It is possible to interpolate the sample reflectivity $\tilde{R}(\nu_t)$ based on known frequency domain information within the ranges $[-\nu_{t\_max}, -\nu_{t\_min}]$ and $[\nu_{t\_min}, \nu_{t\_max}]$, and to obtain low frequency data from the interpolation results. It is convenient to use trigonometric interpolation for this purpose [8],[9]. We apply this type of interpolation for the modulus $|\tilde{R}(\nu_t)|$ and the phase $\varphi\{\tilde{R}(\nu_t)\}$ of the sample complex reflectivity independently. Trigonometric interpolation of the real $Re\{\tilde{R}(\nu_t)\}$ and imaginary $Im\{\tilde{R}(\nu_t)\}$ parts of the sample reflectivity could provide good reconstruction results too. After interpolation the sample pulse function reconstruction take place according to the equation:

$$R^{+}(t) = \mathcal{F}^{-1}\left(\tilde{R}(\nu_t)\right), \tag{2}$$

where $\mathcal{F}^{-1}$ is the inverse Fourier transform operator, $R^{+}(t)$ is the pulse function containing low-frequency components.

To determine the dielectric permittivity profile ε(z) of the sample based on the sample pulse function $R^{+}(t)$ we need to solve the following initial-boundary problem with the sample pulse function as an initial condition:

$$\begin{cases} \dfrac{\partial R(x, s-2x)}{\partial x} = -\dfrac{1}{2}A(x)\displaystyle\int_0^{s-2x} R(x,s')R(x,s-2x-s')ds', \\ \qquad\qquad\qquad\qquad\qquad\qquad\qquad\qquad s > 0, \\ R(1,s) = 0, \qquad\qquad\qquad\qquad\qquad\qquad s > 0, \\ R(x,0) = -\dfrac{1}{4}A(x), \end{cases} \tag{3}$$

where $l = \int_0^L \sqrt{\varepsilon(z)\varepsilon_0\mu_0}dz$ is a time which takes the wavefront to travel through the sample with thickness L, $x = x(z) = \int_0^z \frac{dz'}{lc(z')}$ is a normalized spatial coordinate, $s = \frac{t}{l}$ – is a normalized time coordinate, $R(x,s) = lR^{+}(z,t)$ is a normalized integral kernel corresponding to the sample depth $z(x)$. Note, that $R(0,s) = lR^{+}(t)$ is the initial condition for computations. The function $A(x)$ is a result of the initial-boundary condition (3) numerical solution. Permittivity profile can be easily determined based on $A(x)$ utilizing the following transformation [10]:

$$\begin{cases} z(x) = c(0)l\displaystyle\int_0^x exp\left\{-\int_0^{x'} A(x'')dx''\right\}dx', & 0 < x < 1, \\ \varepsilon\big(z(x)\big) = \varepsilon_1 exp\left\{2\displaystyle\int_0^x A(x')dx'\right\}, & 0 < x < 1. \end{cases} \tag{4}$$

where $c(0)$ is the speed of light in region of space $z < 0$, ε_1 is the permittivity corresponding to the region of space $z < 0$.

The algorithm can be modified to take media absorbance into account. One can use some a priori information about the sample conductivity profile $\sigma(z)$ or some approximation of the conductivity distribution during reconstruction. In addition, it is possible to connect the sample conductivity $\sigma(z)$ with the sample permittivity $\varepsilon(z)$ utilizing a mathematical model of the complex permittivity $\tilde{\varepsilon}$ and Kramers-Kronig relation. In this case, the reconstruction of some $\tilde{\varepsilon}$ parameter profile will take place.

2.2. Experimental results

In order to validate this algorithm, several media with known permittivity profiles were studied. These test media were built from a set of flats with known thickness and THz optical properties. Figure 1 shows an example of the reference and the sample signals detected during the study of the test sample build of 1 *mm* thick quartz window. The results for an algorithm implementation and the priori data

Journal of Physics: Conference Series **486** (2014) 012010 doi:10.1088/1742-6596/486/1/012010

for the sample permittivity is shown in the figure 2 for THz quartz window (a) and the sample built of quartz and HRFZ-SI flats (b).

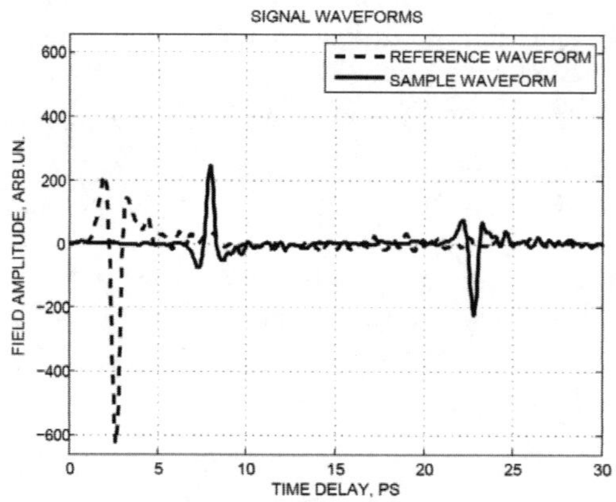

Figure 1. Time-domain spectrometer signals.

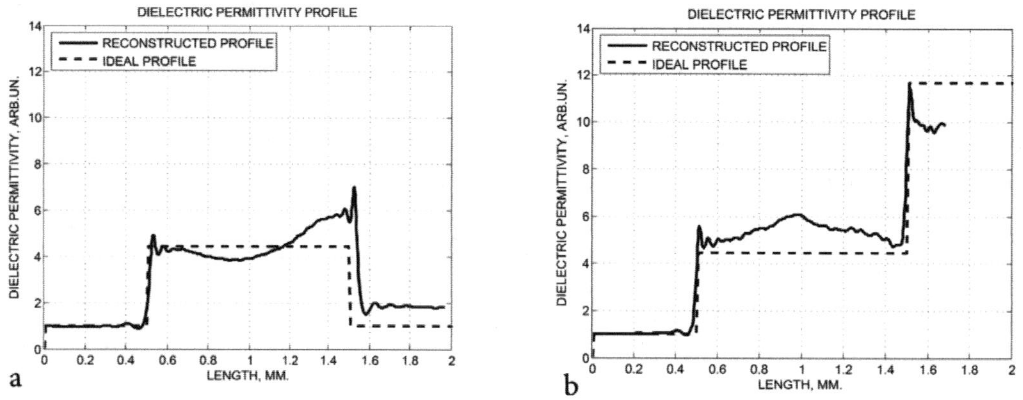

Figure 2. Experimental results.

2.3. Algorithm stability in the presence of white noise

The stability of the permittivity profile reconstruction algorithm was considered by mathematical modelling of the algorithm implementation in the presence of additive white Gaussian noise in TDS system signals. Models of layer medium permittivity profiles as well as models of permittivity profiles with slightly smoothed edges were considered. The interaction of terahertz light with matter was examined utilizing FDTD modelling and reflected waveforms were obtained. White Gaussian noise with various standard deviations was added to the waveforms and the permittivity reconstruction procedure was applied. Results of algorithm implementation were compared with the initial $\varepsilon(z)$ model.

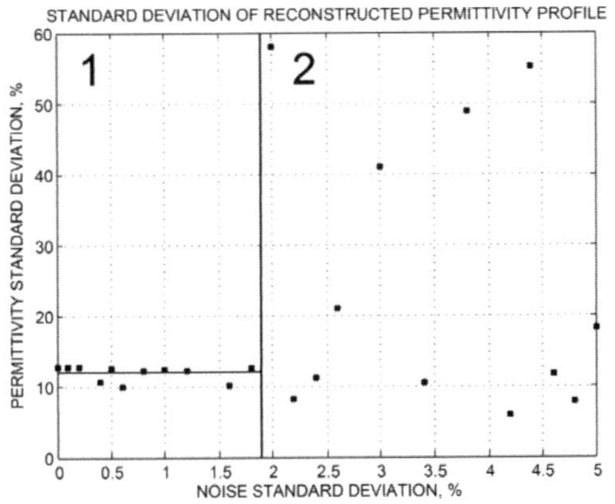

Figure 3. Algorithm stability in the presence of white Gaussian noise.

Dependence of the standard deviation σ_{ε} for the profiles on the waveform noise standard deviation σ_{η} for the same function $\varepsilon(z)$ is shown in figure 3. This figure also contains the border between stable 1) and unstable 2) regions. The dependence of the standard deviation of permittivity profile reconstruction on the noise standard deviation is less than 15.0% for rather high SNR of time-domain data.

3. Conclusions

In the present work, the ability of THz time-domain spectroscopy applied to media dielectric permittivity profile reconstruction was shown. Novel algorithm for solving the inverse scattering problem based on an invariant embedding technique was developed and implemented. The present algorithm can be used in various fields of THz technology, where slightly absorbing media are of interest, such as for non-destructive evaluation of polymer construction materials. This technique will aid the study of internal sample structure.

Acknowledgments

This work was partially supported by grants from the Ministry of Education and Science of Russian Federation (№ 14.B37.21.0898, № 14.B37.21.1282) and from the Russian Foundation for Basic Research (RFBR 12-08-31104; RFBR 12-08-33112, RFBR 14-08-00175, RFBR 14-08-31124).

References

[1] Lee Y-S 2009 *Principles of Terahertz Science and Technology* (Springer, New York, NY)

[2] Federici J F, Schulkin B, Huang F, Gary D, Barat R, Oliveira F and Zimdars D 2005 THz imaging and sensing for security applications - explosives, weapons and drugs *Semicond. Sci. Technol.* **20** 266 – 280

[3] Wallace V P, Woodward R M, Fitzgerald A J, Pickwell E, Pye R J, Arnone D A 2003 Terahertz Pulsed Imaging of Cancers *Proc. SPIE* **4949** 352 – 359

[4] Ashworth P C, Pickwell-MacPherson E, Provenzano E, Pinder S E, Purushotham A D, Pepper M and Wallace V P 2009 Terahertz pulsed spectroscopy of freshly excised human breast cancer *Opt. Express* **17**(15) 12444 – 12454

[5] Stoik C D 2008 *Nondestructive Evaluation of Aircraft Composites using Terahertz Time Domain Spectroscopy* (Ph.D. Thesis, Air Force Institute of Technology*)*

[6] Mittleman D M, Hunsche S, Boivin L, Nuss M C 1997 T-ray tomography *Opt. Lett.* **22**(12) 904–906

[7] Chen Y, Huang S, Pickwell-MacPherson E 2010 Frequency-wavelet domain deconvolution for terahertz reflection imaging and spectroscopy *Opt. Express* **18**(2) 1177 – 1190

[8] Alekhnovich V I, Zaytsev K I, Karasik V E 2013 Medium dielectric permittivity profile reconstruction using terahertz time-domain spectroscopy *Herald of Bauman Moscow State Technical University. Series: "Natural sciences"* **2** 50 – 66

[9] Zaytsev K I, Karasik V E, Fokina I N, Alekhnovich V I 2013 Invariant embedding technique for medium permittivity profile reconstruction using terahertz time-domain spectroscopy *Opt. Eng.* **52**(6) 068203-1 – 068203-10

[10] Kristensson G, Krueger R J 1986 Direct and inverse scattering in the time domain for a dissipative wave equation. I. Scattering operators *J. Math. Phys.* **27** 1667 – 1682

Journal of Physics: Conference Series **486** (2014) 012011
doi:10.1088/1742-6596/486/1/012011

Self-consistent surface charges and electric field in p-i-n tunneling transit-time diodes based on single- and multiple-layer graphene structures

V L Semenenko[1], V G Leiman[1], A V Arsenin[1], V Mitin[2], M Ryzhii[3], T Otsuji[4] and V Ryzhii[5]

[1] Laboratory of Nanooptics and Plasmonics, Moscow Institute of Physics and Technology, Dolgoprudny, Russia
[2] Department of Electrical Engineering, University at Buffalo, Buffalo, U.S.A.
[3] Computational Nanoelectronics Laboratory, University of Aizu, Aizu-Wakamatsu, Japan
[4] Computational Nanoelectronics Laboratory, University of Aizu, Aizu-Wakamatsu, Japan
[5] Research Institute for Electrical Communication, Tohoku University, Sendai, Japan

E-mail: sma_rt2002@mail.ru

Abstract. We develop a device model for p-i-n tunneling transit-time diodes based on graphene single- and multiple-layer structures operating at the reverse bias voltages. The model of the graphene tunneling transit-time diode (GTUNNETT) accounts for the features of the interband tunneling generation of electrons and holes and their ballistic transport in the device i-section, as well as the effect of the self-consistent electric field associated with the self-consistent charges of propagating electrons and holes. Using the developed model, we calculate the dc current-voltage characteristics and the ac frequency-dependent admittance as functions of the GTUNNETT structural parameters, in particular, the number of graphene layers.

1. Introduction

Since the pioneering works [1],[2] on the ballistic electron and hole transport (BET/BHT) in a semiconducting structures, this topic have been attracting an interest of the researchers. Graphene is the perfect material for realization of devices based on the BET for its extremely high achievable charge carriers mobility [3],[4]. In this paper, we consider a p-i-n doped multilayer graphene structure with the applied reverse bias voltage (see Fig. 1a). Electron-hole pairs are generated across the i-section, move in the opposite directions with constant velocities $\pm v_W$, as shown in the figure 1b. Such motion of charges in the i-section (unlike the case of electrons in the 2DEG [5]) leads to the negative AC conductivity of the device in a certain frequency range [6].

At this work, in contrast to the previous treatment [6],[7] we account for the self-consistent electric field associated with the variations of the electron and hole lateral charges in the i-section and their effect on the injection and the dc and ac characteristics (see also V.L. Semenenko et al [8]).

Content from this work may be used under the terms of the Creative Commons Attribution 3.0 licence. Any further distribution of this work must maintain attribution to the author(s) and the title of the work, journal citation and DOI.
Published under licence by IOP Publishing Ltd

2. Results

It is shown that the admittance real part can be negative in a certain frequency range. As revealed, if the i-section somewhat shorter than one micrometer, this range corresponds to the terahertz frequencies. Due to the effect of the self-consistent electric field, the behavior of the GTUNNETT admittance in the range of negativity of its real part is rather sensitive to the relation between the number of graphene layers and dielectric constant. The obtained results demonstrate that GTUNNETTs with optimized structure can be used in efficient terahertz oscillators.

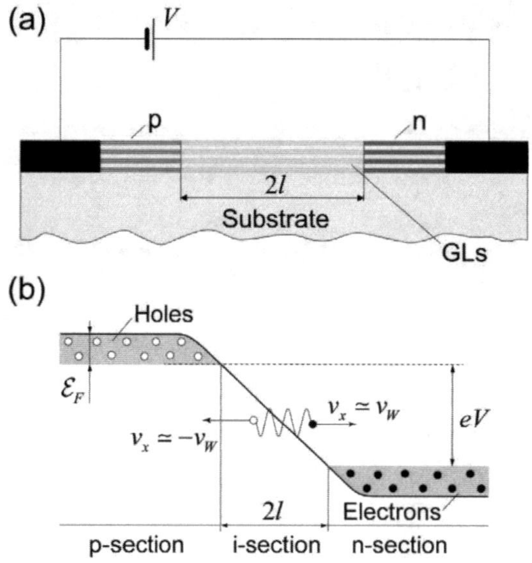

Figure 1. Schematic view (a) of GTUNNETT p-i-n diode with the MGL structure, and (b) its band diagram at reverse bias. Arrows show the propagation directions of electrons and holes generated. due to interband tunneling (mainly in those regions, where the electric field is relatively strong).

References

[1] Shur M S and Eastman L F 1979 *IEEE Trans. Electron Devices* **26** 1677
[2] Shur M S 1981 *IEEE Trans. Electron Devices* **28** 1120
[3] Geim A K and Novoselov K S 2007 *Nat. Mater.* **6** 183
[4] Castro Neto A H, Guinea F, Peres N M R, Novoselov K S and Geim A K 2009 *Rev. Mod. Phys.* **81** 109
[5] Ryzhii V, Satou A and Shur M S 2003 *J. Appl. Phys.* **6**(12) 10041
[6] Ryzhii V, Ryzhii M, Mitin V and Shur M S 2009 *Appl. Phys. Express* **2** 034503
[7] Ryzhii V, Ryzhii M, Shur M S and Mitin V 2010 *Physica E* **42** 719
[8] Semenenko V L, Leiman V G, Arsenin A V, Mitin V, Ryzhii M, Otsuji T and Ryzhii V 2013 *J. Appl. Phys.* **113** 024503

Coherent source terahertz-subterahertz spectroscopy: instrumentation and physics

Boris P Gorshunov and Elena S Zhukova

A. M. Prokhorov General Physics Institute, Russian Academy of Sciences, Russia
Moscow Institute of Physics and Technology (State University), Russia

E-mail: gorshunov@ran.gpi.ru

Abstract. We describe instrumentation and principles of terahertz-subterahertz (THz-subTHz) spectrometers based on continuously frequency tunable coherent radiation sources – backward-wave oscillators (BWOs). The spectrometers cover frequency range from 1 cm^{-1} (0.03 THz) to 50 cm^{-1} (1.5 THz) and operate at temperatures from 2 K to 1000 K, also in magnetic fields up to 8 Tesla. We illustrate abilities of the BWO-spectroscopy by presenting some late results on dielectric, conducting, superconducting and magnetic materials.

1. Introduction and background

More than 20 years ago a way of mastering the THz-subTHz range for spectroscopic purposes was outlined, basing on coherent, continuously frequency-tunable sources, BWOs, and on a quasioptical manner of handling the radiation in measurement schemes [1]. Modern BWO-spectrometers allow for spectroscopic characterization of solid, liquid or gaseous samples, they have made it possible to close the so-called "spectroscopic gap" or the "Terahertz gap" – frequencies 10^{11} Hz < ν < 10^{12} Hz.

2. Instrumentation

The spectroscopic principle is based on measurement of the spectra of two parameters - the transmission coefficient T_r of a plane-parallel sample and the phase shift φ of the radiation passed through this sample (Figure 1).

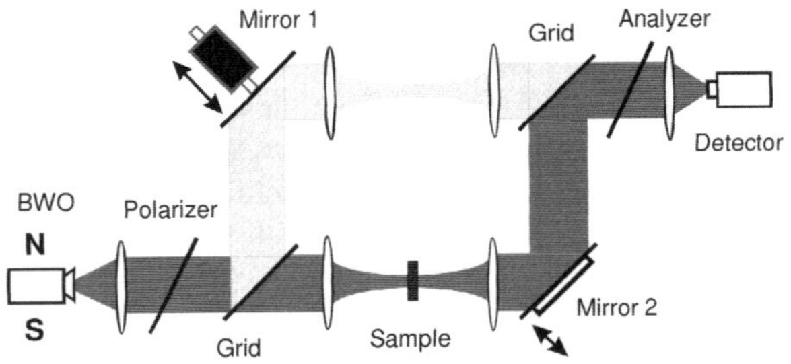

Figure 1. Sketch of Mach-Zehnder interferometric measurement scheme of quasioptical coherent source terahertz-sub-terahertz BWO-spectrometer.

Content from this work may be used under the terms of the Creative Commons Attribution 3.0 licence. Any further distribution of this work must maintain attribution to the author(s) and the title of the work, journal citation and DOI.
Published under licence by IOP Publishing Ltd

Then any pair of "optical" quantities is directly calculated using Fresnel expressions for a complex transmissivity of a plane-parallel slab. The T_r spectrum is measured using the red-colored part of the diagram; to measure the φ, the Mach-Zehnder interferometer is used with the yellow-colored reference arm in operation. The movable mirror (1) is controlled by a software and is electronically tuned to the position providing minimal interferometric signal at the detector. Polyethylene lenses and wire grid polarizers/analyzers are used to manipulate the radiation. The detector is either a Golay cell or a cooled Ge bolometer.

3. Physics

Figure 2 presents typical result – the spectra of cuprate superconductor, Sm-based LSCO single crystal where we have for the first time registered a specific plasma excitation - a transverse Josephson plasmon [2]. A brief overview on THz-subTHz spectroscopy of other electronically correlated systems is given in [3].

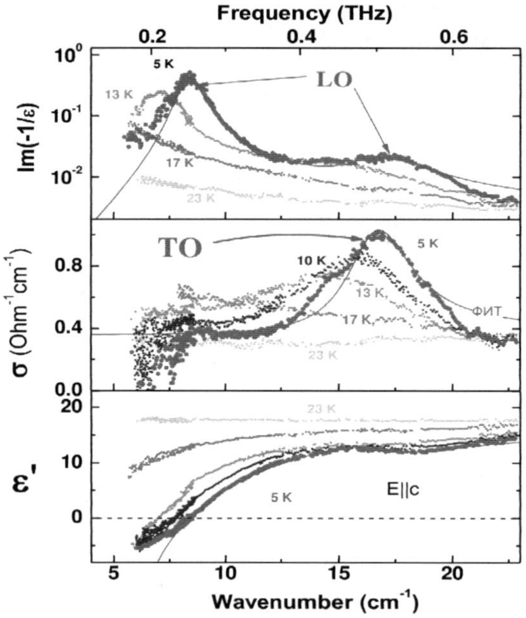

Figure 2. Spectra of loss-function, conductivity and permittivity of $SmLa_{0.85}Sr_{0.15}CuO_{4-\delta}$ crystal. In the normal state, they are typically metallic. In the superconducting state, longitudinal (LO) plasmons develop in the loss function and a new type of excitation, the transverse Josephson plasmon (TO) is observed in the conductivity spectrum [2].

References

[1] Kozlov G V, Volkov A A 1998 Coherent Source Submillimeter Wave Spectroscopy *Top. Appl. Phys.* **74** 51

[2] Kakeshita T, Uchida S, Kojima K M, Adachi S, Tajima S, Gorshunov B and Dressel M 2001 Transverse Josephson plasma mode in T* cuprate superconductors *Phys. Rev. Lett.* **86** 4140

[3] Gorshunov B, Volkov A, Spektor I, Prokhorov A, Mukhin A, Dressel M, Uchida S, Loidl A 2005 Terahertz BWO-spectroscopy *Int. J. Infrared and Millimeter Waves* **26** 1217

Journal of Physics: Conference Series **486** (2014) 012013 doi:10.1088/1742-6596/486/1/012013

Investigation of wide-aperture plasmonic detectors by a tightly focused terahertz beam

D M Yermolaev[1], Ye A Polushkin[1], V E Zemlyakov[1], N A Maleev[2], V V Popov,[3] A V Muravjov[4], S L Rumyantsev[4], M S Shur[4] and S Yu Shapoval[1]

[1]Institute of Microelectronic Technology and High-Purity Materials, Russia
[2]Ioffe Physical Technical Institute, Russia
[3]Kotelnikov Institute of Radio Engineering and Electronics (Saratov Branch), Russia
[4]Rensselaer Polytechnic Institute, USA

E-mail: shapoval@iptm.ru

Abstract. Terahertz response of wide-aperture plasmonic detectors is studied experimentally by using a focused terahertz radiation with frequencies 1.63, 1.89 and 2.55 THz. Two different types of plasmonic detector have been investigated: (i) field-effect transistor (FET) with a grating-gate of large (2×2 mm2) area and (ii) FET with a single gate and ohmic contacts forming a bow-tie antenna. By raster scanning the detector area by using a tightly focused terahertz beam, we studied the contribution of individual parts of the detector to the total detection response and we determined the effective area of the detector structures.

1. Introduction

Up to date, a number of terahertz (THz) plasmonic detectors based on the FET structures have been proposed and investigated [1]-[4]. Most studies have been performed with using the experimental setup where the size of the detector was smaller than the size of the cross section of THz beam. In such experiments, the important question of an effective area of the detector in terms of the detector responsivity and its coupling strength to incident THz radiation could not be experimentally analyzed.

In the present study were investigate two conceptually different THz plasmonic detectors: (i) FET with a single gate and a broadband bow-tie antenna and (ii), the FET structure with a grating-gate of large area.

2. Sample fabrication and measurements

A material system for the devices is an AlGaAs/InGaAs/GaAs heterostructure (see [2]) grown by the molecular beam epitaxy. The top view of grating-gate and bow-tie detectors is shown in figure 1. The microchip holder with the detector was mounted on the translation stage which could move in three dimensions. A continuous source of linearly polarized THz radiation was a gas laser SIFIR-50. The operation parameters of the THz laser are listed in the table 1.

Content from this work may be used under the terms of the Creative Commons Attribution 3.0 licence. Any further distribution of this work must maintain attribution to the author(s) and the title of the work, journal citation and DOI.
Published under licence by IOP Publishing Ltd

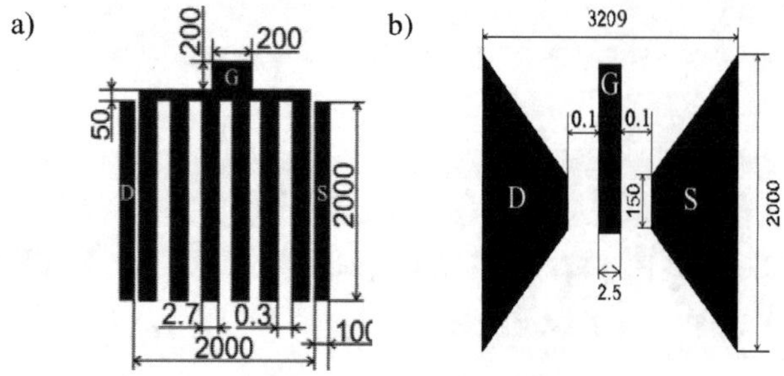

Figure 1. Top view of the investigated detectors (all dimensions are given in microns): a) grating-gate detector; b) bow-tie detector.

Table 1. Operation parameters of THz source.

f, THz	1.63	1.89	2.55
λ, μm	184	159	118
Sspot, mm2	0.027	0.020	0.011
P, mW	143.1	86.6	108.8

In table 1, the following designations are used: f is the operation frequency, λ is the wavelength of THz radiation, S_{spot} is the minimum area of the THz beam spot, $S_{pot}=\pi\times(\lambda/2)^2$, and P is the THz power.

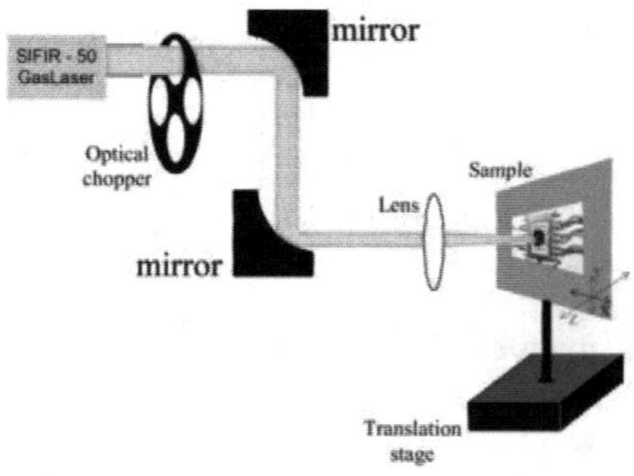

Figure 2. Measurement setup.

Terahertz radiation was chopped at frequency of about 100 Hz, and then, through the focusing optical system, was impinged onto the sample. The radiation beam was focused to a spot with a diameter about the wavelength of the incident radiation by means of parabolic mirrors (figure 2). This allowed us to scan the area of the wide-aperture detector. All measurements were performed at room temperature. Photoresponse signal was measured using a standard lock-in detection technique.

3. Results and discussion

3.1. Bow-tie detector

Raster scans for the bow-tie plasmonic detector at frequencies 1.63 and 2.55 THz are shown in figure 3.

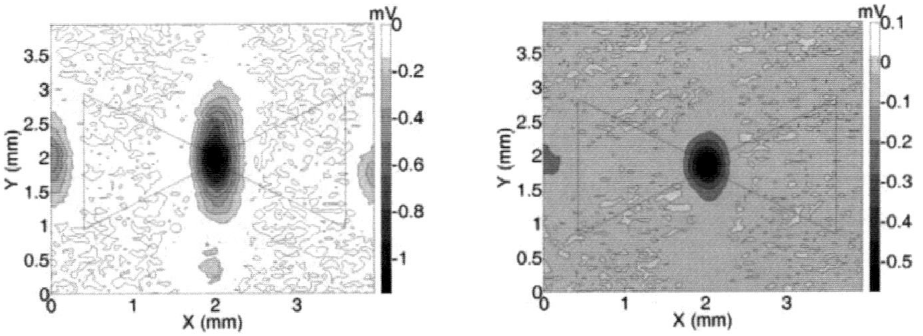

(a) (b)

Figure 3. Raster scans images of the THz photoresponse for the bow-tie plasmonic detector at frequencies (a) 1.63 and (b) 2.55 THz for the drain current $I_d = 25$ μA, gate voltage $V_g = -40$ mV, and the leakage current $I_g = -1.65$ nA. The position of the detector is also schematically shown.

It follows from the raster scan images that the effective detection area is 0.11 and 0.07 mm^2 for 1.63 and 2.55 THz, respectively, with respective responsivities 0.19 V/W and 0.22 V/W. Responsivity, *R*, was calculated by the formula

$$R = \frac{\iint_{image} PhResp \times dxdy}{P \times S_{spot}}, \tag{1}$$

where PhResp is the photoresponse. Double integral is calculated over the image area. The effective image area is calculated at $1/\sqrt{2}$ of the maximum photoresponse in the raster scan image.

3.2. Grating-gate detector

Previously, the grating-gate FET detector was examined for resonance detection of sub-THz radiation in [2], [4]. In those studies, incident THz radiation was uniformly distributed over the area of the sample (the cross-section area of THz beam exceeded the sample area). However, due to possible non-uniformity of a large-area grating-gate structure, different parts on the structure might contribute differently to the overall THz detection response. Scanning the device area by a narrow THz beam, allows us to detect the effective area of THz detection.

Raster scan THz images of the grating-gate detector measured at frequency 1.89 THz for zero drain bias current are shown in figures 4 and 5. Due to the symmetry of the grating gate, there is no detection contribution from the grating area of the structure. It can be seen in figure 4 that the detection contributions from the source and drain contacts, being almost of the value, are of opposite signs. Because of that the contact contributions effectively cancel each other and hence do not virtually affect the overall THz photoresponse. The raster scan THz images of the grating-gate detector for different values of the drain bias current are shown in figure 5.

Journal of Physics: Conference Series **486** (2014) 012013 doi:10.1088/1742-6596/486/1/012013

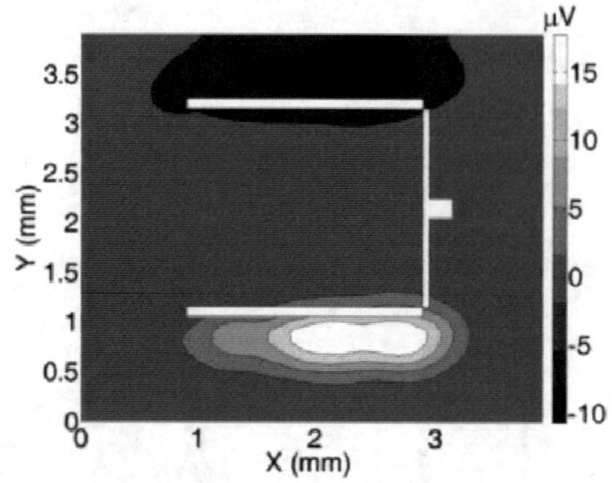

Figure 4. Raster scan images of THz photoresponse of the grating-gate detector for zero gate voltage and $I_d = 0$. The position of the detector is also schematically shown.

Detector responsivities calculated by Eq. (1) are 29, 72, 94, and 155 mV/W for figures from 5a to 5d, respectively. The value of THz photoresponse increases with increasing the bias current. However, the THz photoresponse from the grating-gate area is exhibited even for zero drain bias current (see figure 5c). This can be attributed to a relatively strong gate-leakage current I_g flowing in the channel even for vanishing drain current. Because the leakage current flowing in the source-gate circuit [5] is unidirectional, the photoresponse is different for opposite polarities of the drain bias current (cf. figures 5a and 5d).

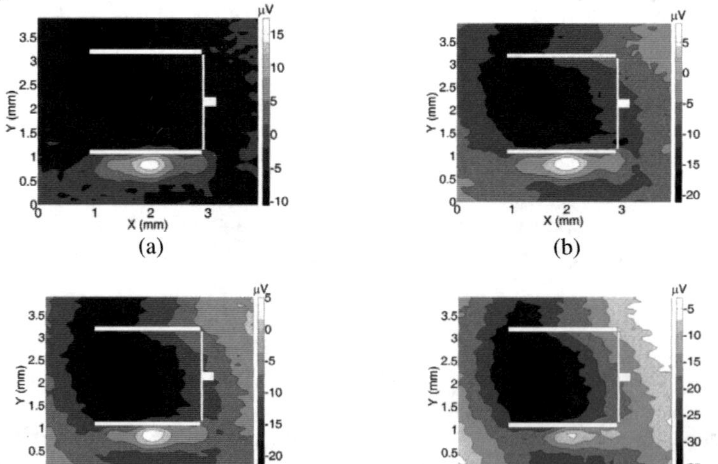

Figure 5. Raster scan images of THz photoresponse of the grating-gate detector for the gate voltage –1 V: a) $I_d = 448$ μA, $I_g = 607$ μA; b) $I_d = 170$ μA, $I_g = 481$ μA; c) $I_d = 0$, $I_g = 412$ μA ; d) $I_d = -448$ μA, $I_g = 268$ μA. The position of the detector is also schematically shown.

It is seen in figure 5 that the effective area of detection is 3.2 mm^2, which is about 80% of the total area of the grating-gate. The decrease in the effective area of the detector is due to the shunting of the THz electric field by the metal bus lead connecting the grating-gate fingers (right white line of the grating-gate contour in figures 5 a-d). This lead affects the distribution of the external incident field over the distance of $2\times\lambda=0.318$ mm from the lead, thereby reducing the effective detection area by 16%. This value is in good correspondence with the measured effective area of the detector (20%). Since the almost entire area of the grating gate along its periodicity retains its effectiveness, it can be concluded that the majority of the grating-gate fingers work effectively.

4. Conclusions

We report on an experimental investigation of the wide-aperture THz plasmonic detectors by using a tightly focused THz beam. This study allows for direct measurement of the effective area of the detector that provides opportunities for further improving the design of large-aperture detectors and can be used for analyzing the coupling efficiency between the detector and THz radiation. This analysis can also form the basis for the development of the quality control of multi-gated FET structures.

Acknowledgement

This work was partly supported by the RAS Programs “Fundamentals and Technological Development of the Nanoelectronic and Nanostructure Sensor Components in the Microwave and Terahertz Ranges” and “Fundamentals of Nanostructure Technologies and Nanomaterials.”

References

[1] Peralta X G, Allen S J , Wanke M C, Harff N E, Simmons J A, Lilly M P, Reno J L, Burke P J, and Eisenstein J P 2002 *Appl.Phys. Lett.* **81** 1627
[2] Yermolaev D M, Marem’yanin K M, Fateev D V, Morozov S V, Maleev N A, Zemlyakov V E, Gavrilenko V I, Shapoval S Yu, Sizov F F and Popov V V 2013 *Solid-State Electron.* **86** 64
[3] Lisauskas A, Pfeiffer U,Öjefors E, Haring Bolìvar P, Glaab D and Roskos H G 2009 *J. Appl. Phys.* **105** 114511
[4] Marem'yanin K V, Ermolaev D M, Fateev D V, Morozov S V, Maleev N A, Zemlyakov V E, Gavrilenko V I, Popov V V, and Shapoval S Yu 2010 *Techn.Phys.Lett.* **36**(4) 365
[5] Dyakonov M and Shur M 1993 *Phys.Rev.Lett.* **71** 15

Journal of Physics: Conference Series **486** (2014) 012014 doi:10.1088/1742-6596/486/1/012014

Medical diagnostics using terahertz pulsed spectroscopy

Kirill I Zaytsev[1], Konstantin G Kudrin[1,2], Svetlana A Koroleva[3], Irina N Fokina[1], Svetlana I Volodarskaya[1], Ekaterina V Novitskaya[1], Artem N Perov[1], Valeriy E Karasik[1] and Stanislav O Yurchenko[1]

[1] Bauman Moscow State Technical University, 2nd Baumanskaya str., 5, Moscow, 105005, Russia
[2] P. A. Gertsen Moscow Research Institute of Oncology, Moscow, Russia
[3] Municipal health institution "Central Regional Hospital of Pavlovskiy Posad", Pavlovskiy Posad, Moscow region, Russia

E-mail: kirzay@gmail.com

Abstract. The paper contains recent results of studying the ability of human body disease diagnosis with terahertz time-domain spectroscopy. *In vitro* skin cancer samples (squamous cell carcinoma, epithelioid cell melanoma, infiltrating carcinoma) were studied experimentally with terahertz pulsed spectrometer. The parametrical *in vitro* images of skin cancers are presented. The ability to make early tooth cariosity diagnosis with terahertz time-domain spectroscopy was also shown experimentally. The results of studying the *in vitro* tooth samples are presented and discussed.

1. Introduction and background

Terahertz (THZ) technology could become an effective tool for medical diagnostics. THz rays are strongly absorbed by water molecules. Human body penetration depth of THz radiation is low due to high water concentration in tissues [1]. But penetration depth is high enough to make a diagnostics of human epithelial tissue diseases, human tooth diseases.

List of papers are dedicated to the medical applications of THz spectroscopy. The potentiality for THz time-domain spectroscopy (TDS) usage for diagnostics of basal cell carcinoma of skin was shown in works [2],[3]. The papers [4],[5] indicate that THz pulsed spectroscopy could become a powerful tool for intraoperative diagnosis of human breast cancer. TDS system would help to find the region of cancer disease during the operation of cancer excision. Also THz spectroscopy could be useful for diagnosis of colon tissue cancers [6], but the problem of THz beam delivery into human colon should be solved at first. The potential application of THz sensing to the field of ophthalmology was considered in paper [7].

2. Results

2.1. Screening diagnostics of skin cancers

There are differences between structure and chemical composition of healthy and diseased tissue which leads to differences in THz spectral reflectivity. Sensitivity of THz radiation to water concentration in tissue allows us to detect cancer on its initial stage.

In the figures 1, 2, 3 the results of different skin cancer studying are presented. The experiments were produced using THz TDS system equipped with the mechanical raster scanning system. The

Content from this work may be used under the terms of the Creative Commons Attribution 3.0 licence. Any further distribution of this work must maintain attribution to the author(s) and the title of the work, journal citation and DOI.
Published under licence by IOP Publishing Ltd

samples were studied *in vitro* in reflection geometry. Parametrical images of skin samples were obtained after processing of collected spectroscopic measurements of raster scan utilizing the method of spectra integration suggested in [8].

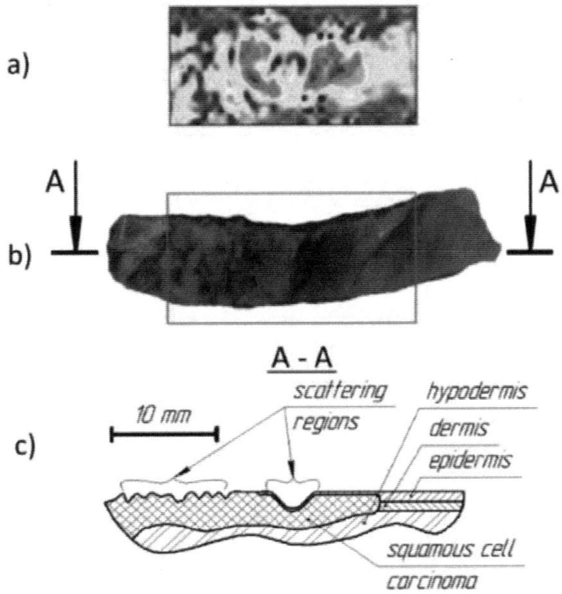

Figure 1. Squamous cell carcinoma:
(a) is a parametrical image, (b) is a photo of *in vitro* diseased skin,
(c) is an illustration of the sample surface profile.

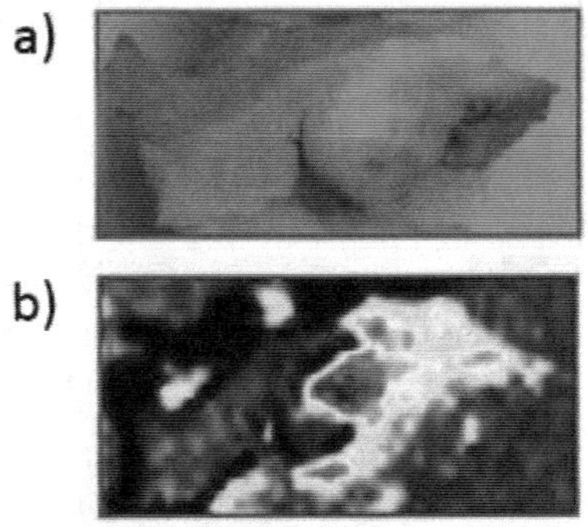

Figure 2. Infiltrating carcinoma: (a) is a photo of *in vitro* diseased skin,
(b) is a parametrical image of the cancer.

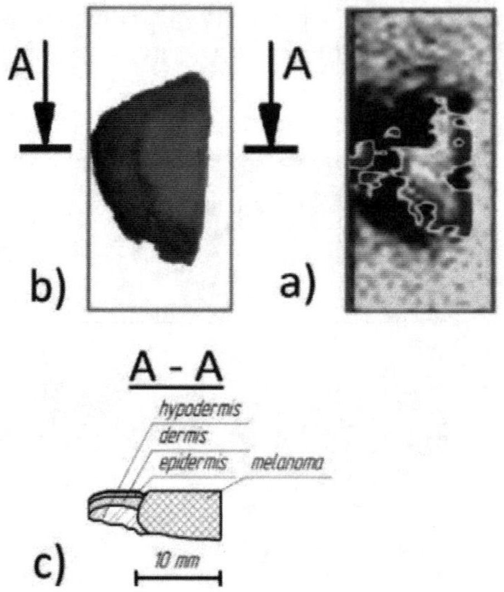

Figure 3. Epithelioid cell melanoma: (a) is a parametrical image, (b) is a photo of *in vitro* diseased skin, (c) is an illustration of the sample surface profile.

Differences in THz spectra of healthy and diseased skin lead to appearance of contrast between healthy skin and sick tissues in parametrical images. High contrast in parametrical images was observed for all examined types of skin cancers. Contrast also presents at the regions of the samples where the cancer lies at a small depth under the layer of the healthy skin (about $\mathbf{0.5}\ \boldsymbol{mm}$).

Figure 4 shows the results of mathematical modelling of parametrical image generation process with TDS system [10]. THz radiation interaction with a diseased skin (basal cell carcinoma) was modelled by means of numerical solving of Maxwell's equations with finite-difference time-domain (FDTD) method. Skin cancers located at a different depth under the healthy epidermis were considered. The modelling results allow us to conclude that the skin cancer diagnosis is possible if the depth of cancer is smaller then $\mathbf{1}\ \boldsymbol{mm}$. Cancer regions with lateral size smaller than $\mathbf{0.5}\ \boldsymbol{mm}$ could be detected. Thus TDS systems and THz imaging systems seem to be a good instrument for the skin cancer screening diagnosis.

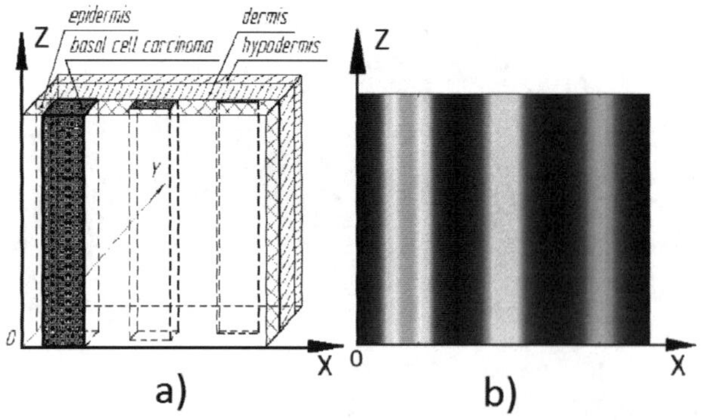

Figure 4. Modeling of the parametrical imaging:
(a) is a model of skin diseased with basal cell carcinoma,
(b) is a results of parametrical image generation.

2.2. Early diagnostics of tooth curiosity.

Low water content in tooth enamel and sensitivity of THz radiation to mineral content in dental tissue make it possible to use THz radiation for tooth demineralization diagnostics (diagnostics of early tooth cariosity) [9]. Mineral content in enamel tissue is strongly related with enamel reflectivity at THz frequencies.

In the figure 5 the result of experimental studying of tooth enamel reflectivity in different demineralization conditions are presented. The stages of demineralization were modelled by means of washing out tooth minerals with orthophosphoric acid (35%). At all stages of minerals leaching the tooth looked healthy and did not contain any white spots. But THz reflectivity of sample differs for all demineralization stages.

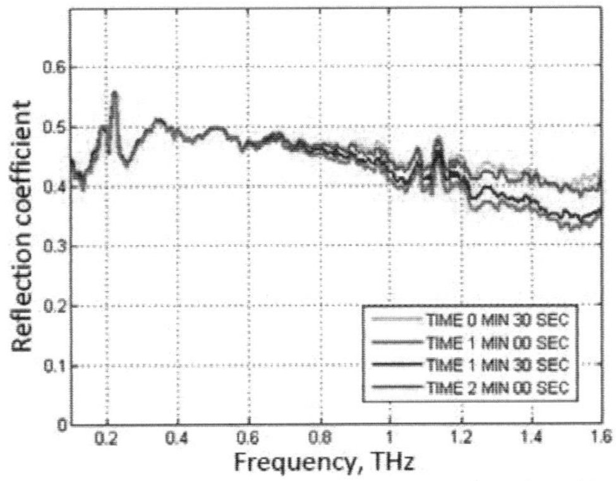

Figure 5. Reflectivity of tooth enamel *in vitro* after leaching of minerals with orthophosphoric acid (35%) for **30 *sec.***, **1 *min.***, **1 *min.* 30 *sec.***, **2 *min.***

Notice, the spectral reflectivity is decreasing during the enamel demineralization especially at high frequencies. It could help us to make early tooth cariosity diagnostics with TDS system [11].

Conclusions

The present paper has gave the brief overview on biomedical applications of THz pulsed spectroscopy. Recent results of studying the ability to make early diagnosis of skin cancers and tooth enamel demineralization with THz radiation were presented and discussed.

Acknowledgements

This work was partially supported by grants from the Russian Foundation for Basic Research (RFBR 14-08-31102; RFBR 14-02-00781).

References

[1] Bunkin N F, Yurchenko S O, Suyazov N V, Shkirin A V 2012 Structure of the nanobubble clusters of dissolved air in liquid media *J. Biol. Phys.* **38**(1) 121-152

[2] Brun M-A, Formanek F 2010 Terahertz Imaging Applied to Cancer Diagnosis *Physics in Medicine and Biology* **55**(16) 4615 – 4623

[3] Wallace V P, Woodward R M, Fitzgerald A J, Pickwell E, Pye R J, Arnone D A 2003 Terahertz Pulsed Imaging of Cancers *Proc. SPIE* **4949** 352 – 359

[4] Fitzgerald A J, Wallace V P 2006 Terahertz Pulsed Imaging of Human Breast Tumors *Radiology* **239**(2) 533 – 540

[5] Ashworth P C, Pickwell-MacPherson E, Provenzano E, Pinder S E, Purushotham A D, Pepper M, Wallace V P 2009 Terahertz pulsed spectroscopy of freshly excised human breast cancer *Opt. Express* **17**(15) 12444–12454

[6] Reid C. 2009 *Spectroscopic Methods for Medical Diagnosis at Terahertz Wavelength* (Ph. D. Thesis, University College London)

[7] Bennett D B, Taylor Z D, Tewari P, Singh R S, Culjat M O, et. al 2011 Terahertz sensing in corneal tissues *J. Biomed. Opt.* **16**(5) 057003-1 – 057003-8

[8] Zaytsev K I, Karasik V E, Koroleva S A, Fokina I N 2012 Study the ability of terahertz imaging system application for medical diagnosis *Herald of Bauman Moscow State Technical University. Series: "Instrumentation"* **4** 114 – 127

[9] Zaytsev K I, Karasik V E, Novitskaya E V, Ryzhii V I, Fokina I N 2012 Experimental studying the ability for skin cancer diagnosis with terahertz spectroscopy *Herald of Bauman Moscow State Technical University. Series: "Instrumentation"* **S9** 20 – 32

[10] Pickwell E, Wallace V P, Cole B E, Ali S, Longbottom C, Lynch R J M, Pepper M 2007 A Comparison of Terahertz Pulsed Imaging with Transmission Microradiography for Depth Measurement of Enamel Demineralisation in vitro *Caries Research* **41** 49–55

[11] Volodarskaya S I, Zaytsev K I, Karasik V E, Novitskaya E V, Fokina I N 2012 Experimental studying the ability of tooth enamel early cariosity diagnosis with terahertz spectroscopy *Herald of Bauman Moscow State Technical University. Series: "Instrumentation"* **S9** 33 – 48

Journal of Physics: Conference Series **486** (2014) 012015 doi:10.1088/1742-6596/486/1/012015

Terahertz probing of local electron states in $Pb_{1-x}Sn_xTe(In)$

Vladimir Chernichkin[1], Ludmila Ryabova[1], Andrey Nicorici[2], Sergey Danilov[3] and Dmitry Khokhlov[1]

[1] M. V. Lomonosov Moscow State University, Russia
[2] Institute of Applied Physics, Kishinev, Moldova
[3] University of Regensburg, Germany

E-mail: khokhlov@mig.phys.msu.ru

Abstract. A new type of semiconductor local states is revealed in lead-tin telluride solid solutions doped with indium. The energy position of these states is not linked to any specific location in the semiconductor energy spectrum, but follows the quasiFermi level position, which may be tuned by photoexcitation. The binding energy of these states is less than 10 meV providing appearance of photoresponse at wavelengths exceeding 100 μm. This conclusion is based on the results of experiments that use illumination of the samples by 100 ns – long pulses of an optically pumped NH_3 terahertz laser with the wavelengths of 90, 148 and 280 μm corresponding to the quanta energies of 13.8, 8.4 and 4.4 meV, respectively.

1. Introduction and background

Doping of $Pb_{1-x}Sn_xTe$ with indium results in pinning of the Fermi level in a position that is defined only by the mole fraction of SnTe x. This position of the indium deep ground impurity state provides the Fermi level pinning. Its location may be tuned from ~ 70 meV above the conduction band edge in PbTe(In) providing free electron concentration of 7 10^{18} cm^{-3}, all the way through the gap at $0.22 < x < 0.28$ giving rise to a semiinsulating state at low temperatures, down to ~ 15 meV below the valence band edge in $Pb_{0.7}Sn_{0.3}Te(In)$ which leads to appearance of free holes. Besides that, the persistent photoconductivity effect observed in this semiconductor system at low temperatures $T < 25$ K allows variation of the quasiFermi level position by photoexcitation.

2. Results

We investigated photoconductivity in single crystals of $Pb_{1-x}Sn_xTe(In)$ with x varied within 0.22-0.29, and in highly degenerated polycrystalline PbTe(In) films. All experiments with laser pulses were performed without screening samples from background radiation. The persistent photoconductivity induced by this background radiation has resulted in generation of long-lived non-equilibrium electrons at low temperatures and, consequently, to formation of a quasiFermi level in the conduction band. The distance between the quasiFermi level and conduction band bottom depended on the alloy composition and reached its maximum of ~ 100 meV in PbTe(In) films. Nevertheless a positive persistent photoresponse to the action of laser pulses was observed for all investigated samples at $T < 25$ K. The laser quanta energy was significantly lower than the energy separation between the quasi Fermi level and the conduction band edge or the ground impurity state. Tuning of the quasiFermi level position in PbTe(In) by means of an additional optical pumping of long-lived non-equilibrium electrons allowed us to reduce conductivity by 20% of its initial value and to push the

Content from this work may be used under the terms of the Creative Commons Attribution 3.0 licence. Any further distribution of this work must maintain attribution to the author(s) and the title of the work, journal citation and DOI.
Published under licence by IOP Publishing Ltd

quasiFermi level position up by ~ 20 meV. No change of the relative photoresponse was observed for all of the laser wavelengths (figure1). It means that the local electron states remain in the close vicinity of the quasi Fermi level even after it moves to another position. It approves our suggestion that the terahertz photoresponse we observe is due to the excitation from the local states linked to the quasi Fermi level.

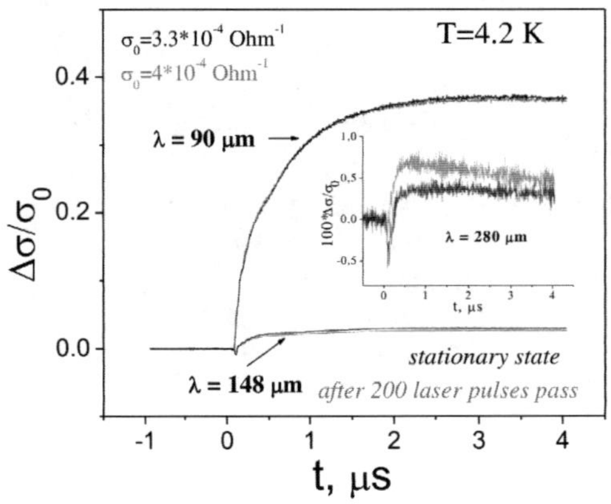

Figure 1. Kinetics of the photoconductivity of a PbTe(In) film at different laser wavelengths and at different levels of photoexcitation.

Dependence of the photoresponse amplitude as a function of magnetic field has been studied in $Pb_{0.75}Sn_{0.25}Te(In)$. It is demonstrated that this dependence has a pronounced peak. The position of this peak in magnetic field is proportional to the frequency of the laser radiation (figure 2).

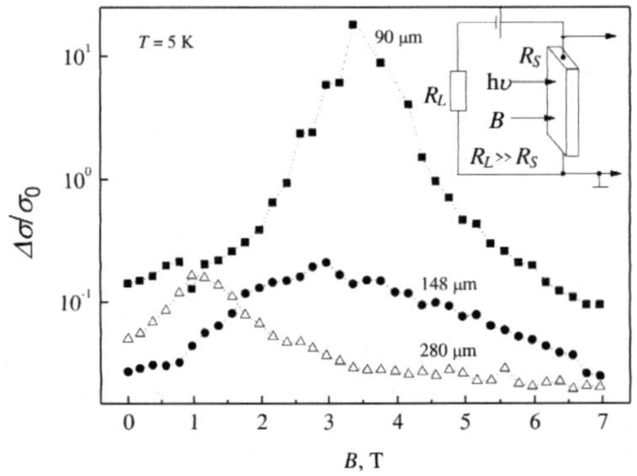

Figure 2. Amplitude of the persistent photoconductivity in $Pb_{0.75}Sn_{0.25}Te(In)$ as a function of magnetic field at different laser wavelengths. The experiment geometry is shown in the insert.

Terahertz detector with series connection of asymmetric gated transistors

D M Yermolaev[1], K M Marem'yanin[2], N A Maleev[4], V E Zemlyakov[1], V I Gavrilenko[2], V V Popov[3] and S Yu Shapoval[1]

[1] Institute of Microelectronic Technology and High-Purity Materials, Russia
[2] Institute for Physics of Microstructures, Russia
[3] Kotelnikov Institute of Radio Engineering and Electronics (Saratov Branch), Russia
[4] Ioffe Physical Technical Institute, Russia

E-mail: yermolayev.denis@gmail.com

Abstract. Terahertz (THz) detection by a one-dimensional array of series connected field-effect transistors (FETs) is studied experimentally. Such terahertz detector demonstrates greatly enhanced voltaic responsivity up to 2 kV/W. Asymmetrical position of the gate contact in each FET in the array enables strong photovoltaic response.

1. Introduction and background

Recent years have witnessed further progress in THz detection based on plasmonic nonlinearities in a FET with two-dimensional (2D) electron channel, which was originally proposed in [1]. Resonant [2],[3] as well nonresonant [4],[5] plasmonic detectors has been studied.

In Ref. 5, we have shown that a densely ordered 1D array of FETs with asymmetrical position of the gate contact in each FET can be used as a THz detector with enhanced responsivity without using supplementary antenna elements. But we have used a parallel connection of FETs in this work and the enhanced responsivity was related to the single FET in array while the voltaic responsivity of the overall array was relative low. After we made 1D array of series connected such FETs and measured its voltaic responsivity.

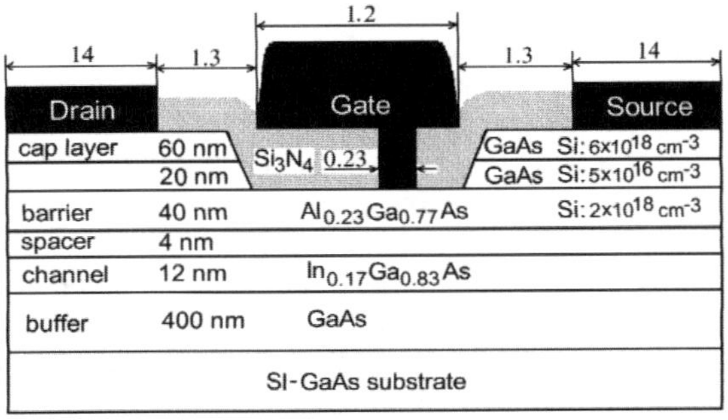

Figure 1. Schematic of an individual FET in the array (all lateral dimensions are given in microns).

Content from this work may be used under the terms of the Creative Commons Attribution 3.0 licence. Any further distribution of this work must maintain attribution to the author(s) and the title of the work, journal citation and DOI.
Published under licence by IOP Publishing Ltd

2. Results

Single FET and transistor's structure are on figure 1 and 2.

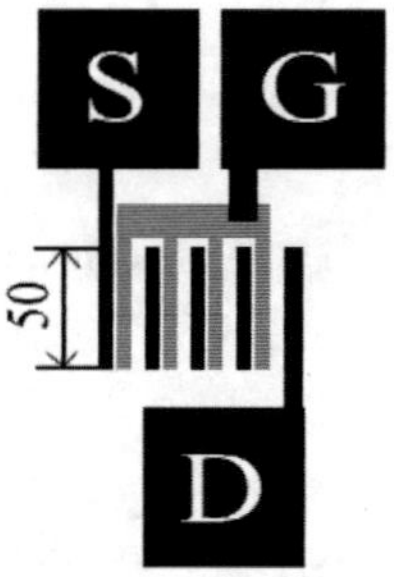

Figure 2. Top view of the detector array.

Measured photoresponse are on figure 3.

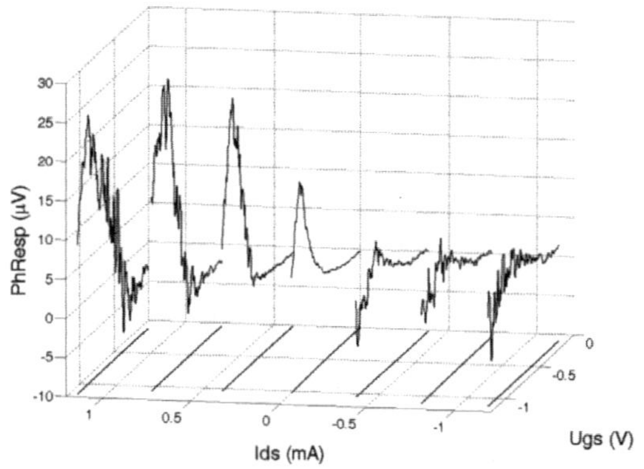

Figure 3. The dependence of the photoresponse on the gate-source voltage and the drain-source current at 0.6 THz (line at the bottom of the chart is the corresponding projections of the photoresponse curves on the coordinate plane XY).

The maximum sensitivity of the photovoltanic response is 1095 V/W. Increased sensitivity by passing the bias current is 1910 V/W at Ids = 0.77 mA. These values are better than into Ref. [6].

References

[1] Dyakonov M and Shur M 1996 *IEEE Trans. Electron Devices* **43** 380

[2] Fatimy A El, Teppe F, Dyakonova N, Knap W, Seliuta D, Valušis G, Shchepetov A, Roelens Y, Bollaert S, Cappy A and Rumyantsev S 2006 *Appl. Phys. Lett.* **89** 131926

[3] Yermolaev D M, Marem'yanin K M, Fateev D V, Morozov S V, Maleev N A, Zemlyakov V E, Gavrilenko V I, Shapoval S Yu, Sizov F F and Popov V V 2013 *Solid State Electronic* **86** 64-67

[4] Tauk R, Teppe F, Boubanga S, Coquillat D, Knap W, Meziani Y M, Gallon C, Boeuf F, Skotnicki T, Fenouillet-Beranger C, Maude D K, Rumyantsev S and Shur M S 2006 *Appl. Phys. Lett.* **89** 253511

[5] Popov V V, Ermolaev D M, Maremyanin K V, Maleev N A, Zemlyakov V E., Gavrilenko V I, Shapoval S Yu 2011 *Appl. Phys. Lett.* **98** 153504

[6] Elkhatib T A, Kachorovskii V Y, Stillman W J, Veksler D B, Salama K N, Zhang X-C and Shur M S 2010 *IEEE Trans. Microwave Theory Tech.* **50**(2)

Terahertz emission from silicon nanostructures heavily doped with boron

Nikolay T Bagraev[1], Eduard Yu Danilovskii[1], Dmitrii S Gets[1], Andrey K Kaveev[2], Leonid E Klyachkin[1], Grigorii I Kropotov[2], Andrey A Kudryavtsev[1], Roman V Kuzmin[1], Anna M Malyarenko[1], Vladimir A Mashkov[1], Ivan A Tsibizov[2], Dmitrii I Tsypishka[2] and Ilya A Vinerov[2]

[1] Ioffe Physical Technical Institute, St. Petersburg, Russia
[2] TYDEX, J.S.Co., Domostroitelnaya str. 16, 194292 St. Petersburg, Russia

E-mail: kav@tydex.ru

Abstract. We present the first findings of the terahertz emission from the ultra-narrow p-type silicon quantum well confined by the δ-barriers heavily doped with boron on the n-type Si (100) surface. The THz spectra revealed by the voltage applied along the Si-QW plane appear to result from the radiation of the dipole boron centers.

1. Introduction and background

The device has been prepared on the n-type Si (100) surface within frameworks of silicon planar technology. Making a mask and performing photolithography after preliminary oxidation, the short time diffusion of boron was used to obtain the ultra-shallow p+-n junctions [1]. The cyclotron resonance measurements as well as the infrared and local tunneling spectroscopy data have shown that the p+-boron diffusion profile represent the ultra-narrow p-type silicon quantum well (Si-QW), $2\,nm$, confined by the δ-barriers, $3\,nm$, heavily doped with boron on the n-type Si (100) surface (see figure 1). The SIMS and STM studies have shown that the δ-barriers heavily doped with boron in the concentration of $5 \cdot 10^{21}\ cm^{-3}$ represent really alternating arrays of silicon empty and doped dots, with dimensions restricted to $2\,nm$. This value of the boron concentration seems to indicate that each doped dot located between empty dots contains two impurity atoms of boron. The EPR studies show that these boron pairs are the trigonal dipole centers, which result from the negative-U reconstruction of the shallow boron acceptors, $2B(0) => B(+) + B(-)$. The electrical resistivity, magnetic susceptibility and specific heat measurements demonstrate that these dipole boron centers inside the δ-barriers appear to create the correlation energy gap which is regenerated in the superconductor energy gap, with the value of $0.044\,eV$ corresponding to the critical temperature and magnetic field equal to $T_c = 145\,K$ and $H_{c2} = 0.22\,T$ [1]. Besides, the correlation energy gap appears to cause the THz emission under the voltage applied along the Si-QW plane, with the frequency selected by varying the dimensions of the microcavities created between contacts (see figure 1).

Content from this work may be used under the terms of the Creative Commons Attribution 3.0 licence. Any further distribution of this work must maintain attribution to the author(s) and the title of the work, journal citation and DOI.
Published under licence by IOP Publishing Ltd

Journal of Physics: Conference Series **486** (2014) 012017 doi:10.1088/1742-6596/486/1/012017

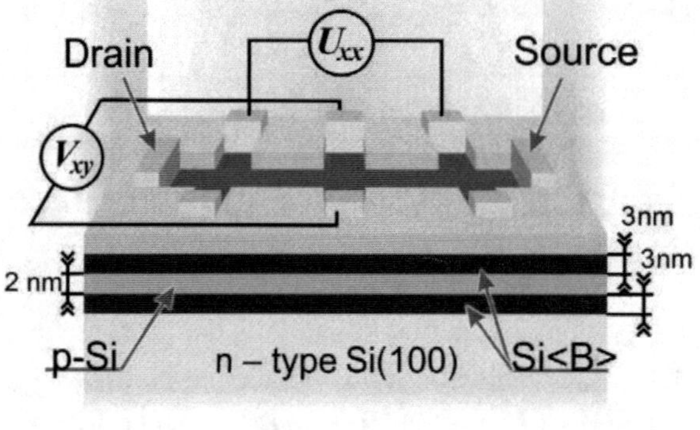

Figure 1. Device schematic, showing the perspective view of the silicon device structure performed within the frameworks of the Hall geometry. The silicon sandwich represents the p-type silicon quantum well confined by the δ-barriers heavily doped with boron on the n-type Si (100) surface. The source and drain Ohmic contacts are marked by S and D, respectively. The terahertz emission is caused by the drain-source current, Ids.

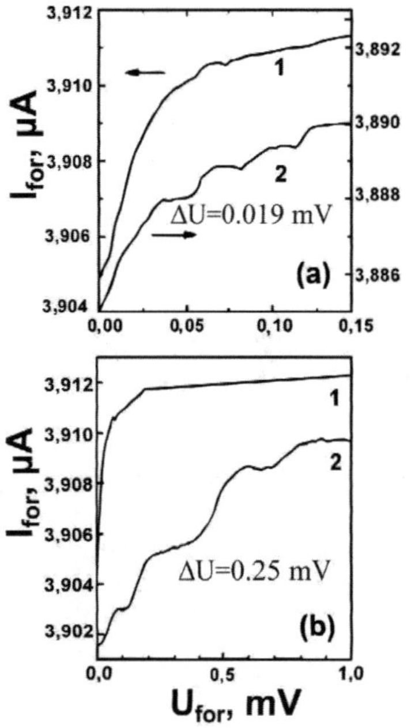

Figure 2. The Shapiro steps that reveal the THz frequency generation at $0.0093\,THz$ (a) and $0.129\,THz$ (b). The drain-source current was stabilised. $h\nu = 2e\Delta U$; ΔU denotes the step length [2]; curves 1 - $I_{ds} = 0$; curves 2 - $I_{ds} = 10\,nA$. $T = 77\,K$.

Journal of Physics: Conference Series **486** (2014) 012017 doi:10.1088/1742-6596/486/1/012017

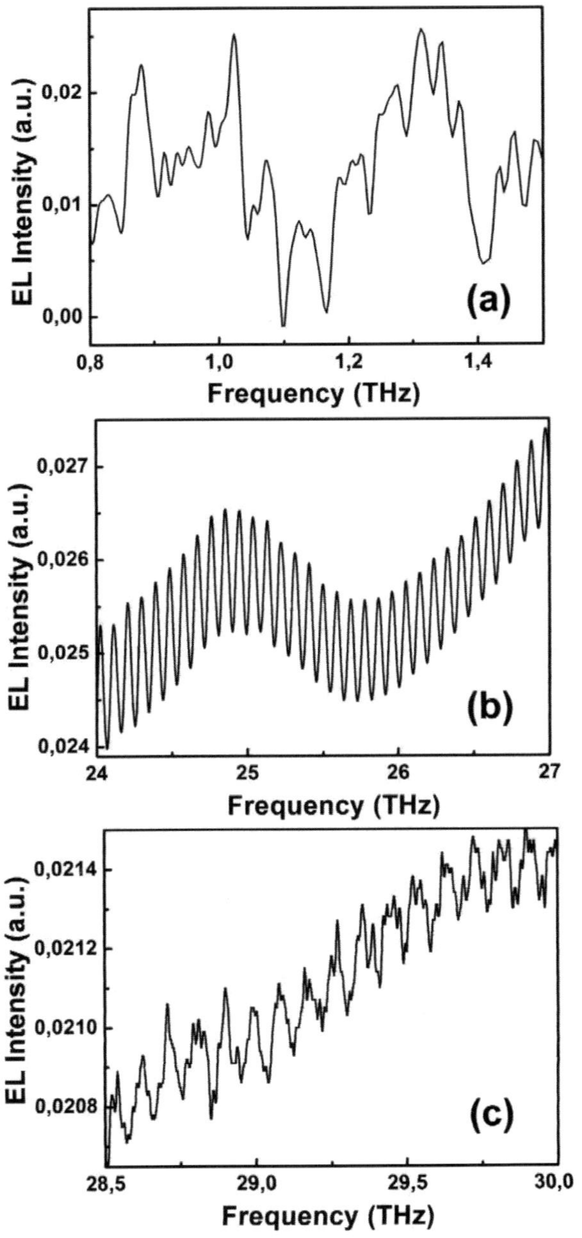

Figure 3. The THz modulated black body radiation that is induced by the stabilized drain-source current in different spectral range; $T=300\ K$; $I_{ds}=40\ mA$; The modulation frequency of 0.129 THz (a, b, c) and 0.0093 THz (a, c) due to the THz emission of the dipole boron centers are present.

2. Results

The THz and GHz emission caused by the radiation of the dipole boron centers was revealed by measuring the Shapiro and Fiske steps under the stabilized drain-source current that defined the values of the frequency, ν, as $0.0093\,THz$ and $0.129\,THz$ which are in a good agreement with the cross and longitudinal sizes, d, of the device, $d = c/2n\nu$; where n denotes refraction coefficient (figure 2a and 2b). The energy subbands of the two-dimensional holes in the Si-QW valence band have been also identified by the optical and electrical measurements.

This THz emission appears to be revealed as the modulation frequency of the black body radiation of the device that was measured with the Bruker-Physik VERTEX 70 FT-IR spectrometer (Figures 3a, b and c). These findings seem to be due to the oscillations of the heat capacity that are induced by the THz emission of the dipole boron centers inside the δ-barriers confining the Si-QW. Finally, by varying the Vxy applied voltage and the drain-source current, the phase shifts of the THz modulation appear to be observed as a result of the spin interference effects (figures 4a and b)

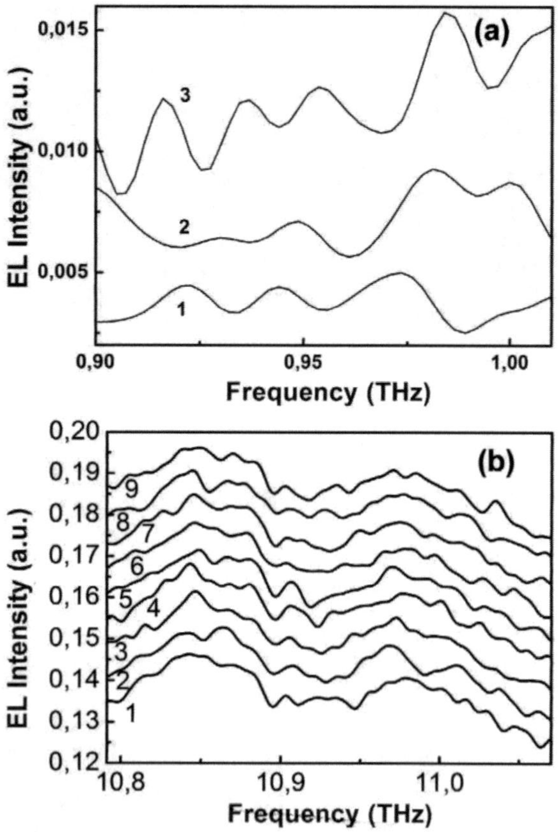

Figure 4. The phase shifts of the $0.129\,THz$ and $0.0093\,THz$ modulation spectra induced by varying the stabilized drain-source current (a) and by biasing the voltage applied to the Hall contacts; $T = 300\,K$; (a) $Ids = 30\,mA$ (1), $40\,mA$ (2), $80\,mA$ (3). (b) $Vxy\,(mV)$: 1 - -250; 2 - -150; 3 - -50; 4 – 0; 5 - $+50$; 6 - $+100$; 7 - $+150$; 8 - $+200$; 9 - $+250$.

References

[1] Bagraev N T, Danilovsky E Yu, Gets D S , Gehlhoff W, Klyachkin L E,. Kudryavtsev A A, Kuzmin R V, Malyarenko A M, Romanov V V 2011 *J. Mod. Phys.* **2** 256
[2] Shapiro S, Janus A R, Holly S 1964 *Rev. Mod. Phys.* **36** 223

Journal of Physics: Conference Series **486** (2014) 012018 doi:10.1088/1742-6596/486/1/012018

Novel Algorithm for Sample Material Parameter Determination using THz Time-Domain Spectrometer Signal Processing

Kirill I Zaytsev[1], Arseniy A Gavdush[1], Sergey P Lebedev[2] and Stanislav O Yurchenko[1]

[1] Bauman Moscow State Technical University, 2nd Baumanskaya str., 5, Moscow, 105005, Russia
[2] A. M. Prokhorov General Physics Institute of the Russian Academy of Sciences, Russia

E-mail: kirzay@gmail.com

Abstract. In our research we have developed a novel sample material parameter determination algorithm which could be used for characterization of very thin flats. Utilizing this algorithm one could find optical properties of the media which are thin enough to cause appearance of the satellite pulses in TDS system signal, transmitted through the flat sample. The algorithm is based on the minimization of the Error function (minimization between the theoretical model of sample transfer function and the experimental data collected with THz pulsed spectrometer). The algorithm was implemented experimentally and tested by means of the test samples characterization. The results of the algorithm implementation were compared with the results of the same test sample studying using terahertz backward-wave oscillator spectroscopy.

1. Introduction and background

Terahertz (THz) time-domain spectroscopy (TDS) is a technique for measuring the complex refractive index (or the complex permittivity) of different media over a wide frequency range starting from 0.1 THz and goes up to 3.0 ... 30.0 THz depending on the methods which are used for generation and detection of short THz pulses. Since the appearance of THz pulsed spectroscopy in second half of XX century [1],[2] a lot of authors have presented their material parameter extraction algorithms working with solid and liquid media [3]-[6].

Usually for material parameter extraction, a THz pulse, transmitted through the sample of interest, is being compared with THz pulse obtained without the sample on the propagation path of THz beam. Flat sample acts as a Fabry-Perot resonator for the THz pulse, thus the measured signal exhibits multiple reflections of the THz pulse (satellite pulses) caused by the multipath interference phenomenon. A simplistic method for approximation of the optical material parameters considers a narrow time window filtering procedure for the sample waveform. The window helps to suppress all the satellite pulses in the sample signal of THz spectrometer, leaving the main sample pulse untouched [7],[8]. However, the isolation of the first pulse may not be possible in the cases of low refractive indexes and/or thin samples Therefore the novel sample material parameter extraction procedure should be developed.

Content from this work may be used under the terms of the Creative Commons Attribution 3.0 licence. Any further distribution of this work must maintain attribution to the author(s) and the title of the work, journal citation and DOI.
Published under licence by IOP Publishing Ltd

Journal of Physics: Conference Series **486** (2014) 012018 doi:10.1088/1742-6596/486/1/012018

The present paper describes a technique for sample material parameter characterization based on the method, described in [5]. The sample could have rather small refractive index and could be rather thin (up to 0.1 mm). Small thickness of the sample would not cause distortion of the reconstruction results. Let's describe all stages of algorithm implementation step by step. The results of experimental studying of test sample will be considered and discussed. Reconstructed sample material parameter curves will be compared with the results of the same sample studying utilizing THz backward-wave oscillator spectroscopy.

2. Results

Sample thickness should be determined precisely using micrometer or Vernier compass before measurement of TDS signals would take place. After thickness determination, two signals are measured with THz pulsed spectrometer: $E_s(t)$ is the signal transmitted through the sample and $E_r(t)$ is the signal transmitted through the empty THz beam path. After detection, the waveforms is being filtered using the procedure of tukey window filtering, which helps to suppress noises caused by multiple reflections of the main THz pulse in THz spectrometer beam path (generator, beamsplitters, lenses, detector etc.). Note, the filtering procedure should not distort the main THz pulse and the satellite pulses caused by the multiple reflections in the sample flat.

Figure 1 contains initial waveforms and the result of filtration. The filter window size and the filter location are determined automatically during the TDS signal processing in time-domain. We have chosen two test samples to verify the algorithm experimentally: $GaAs$ sample with thickness 0.56 mm and $ceramic$ sample with thickness 1.37 mm. The material parameters of $GaAs$ were fully characterized in [9]. Note, that the waveforms presented in the figure 1 correspond to the test $GaAs$ sample.

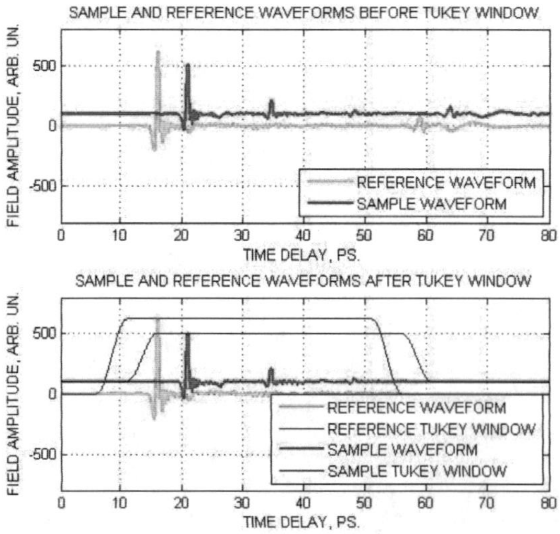

Figure 1. An example of sample and reference waveforms before Tukey window filtering (*top curves*) and after it (*bottom curves*).

Reliable frequencies for complex refractive index determination are lying in range from 0.2 to 2.5 THz for TDS spectrometer based on $LT - GaAs$ photoconductive antenna emitter and $ZnTe$ electrooptical detector. Certainly this restriction depends on the type of TDS system which was used to collect experimental data. After signals have been detected, one could find experimental transfer function of the sample according equation:

$$H_{exp}(f) = \frac{\hat{E}_s(f)}{\hat{E}_r(f)} = \frac{F_t\{E_s(t)\}}{F_t\{E_r(t)\}}, \tag{1}$$

where $\widehat{E}_s(f)$and $\widehat{E}_r(f)$ are Fourier spectrums of the sample $E_s(t)$ and the reference $E_r(t)$ signals, and F_t is a Fourier transform operator. This transfer function takes into consideration all satellite pulses of sample signal.

Quasi Fabry-Perot resonator model was chosen as a theoretical model of sample transfer function. Theoretical model of transfer function has several parameters which need to be determined before starting the material parameter extraction. As it was mentioned above sample thickness l should be known *priori*. Also we need to know the number of satellite pulses in sample signal δ, and the interval of possible values of real and complex part of sample refractive index. The formulation of theoretical model can be written as follows:

$$H_{theor}(f) = 4 \cdot \exp\left(\frac{j \cdot 2\pi f l(\tilde{n}_0 - \tilde{n}_1)}{c_0}\right) \cdot \left(\frac{\tilde{n}_1}{\tilde{n}_1 + \tilde{n}_0}\right)^2 \cdot \\ \cdot \left[1 + \sum_{i=1}^{\delta}\left(\left(\frac{\tilde{n}_0 - \tilde{n}_1}{\tilde{n}_1 + \tilde{n}_0}\right)^2 \cdot \exp\left\{\frac{-2j \cdot 2\pi f \tilde{n}_1 l}{c_0}\right\}\right)^i\right], \tag{2}$$

where f is frequency, c_0 is the speed of light in vacuum, $\tilde{n}_1$ and $\tilde{n}_0$ are the complex refractive index of sample and air, respectively. Figure 2 illustrates the modulus and the phase of the experimental and the theoretical transfer functions.

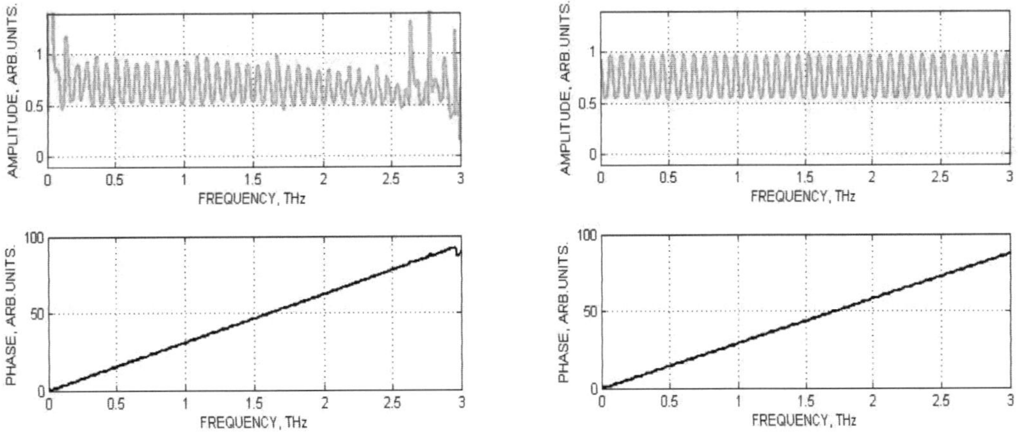

(a) **(b)**

Figure 2. The modulus and the phase curves of the experimental $H_{exp}(f)$ (a) and theoretical $H_{theor}(f)$ (b) transfer functions.

The sample material parameter determination is produced during the minimization of an error functional based on the experimental $H_{exp}(f)$ and theoretical $H_{theor}(f)$ transfer functions:

$$\tilde{n}_1(f) = argmin_{\tilde{n}_1(f)}\{Err(f, \tilde{n}_1(f), l)\} \tag{3}$$

The error functional can be described with the following equation:

$$Err(f, \tilde{n}_1(f), l) = (\xi \cdot |M(f)| + \eta \cdot |A(f)|), \tag{4}$$

where

$$\begin{cases} M(f) = |H_{theor}(f)| - |H_{exp}(f)|, \\ A(f) = \angle H_{theor}(f) - \angle H_{exp}(f), \\ \eta = 10 \cdot \xi. \end{cases} \tag{5}$$

Coefficients η and ξ helps to correct differences in order of the functions $M(f)$ and $A(f)$. In the figure 3 the dependence of an error function on the test *ceramic* sample refraction index, absorption coefficient and on the frequency of the incident electromagnetic wave is shown.

Inaccurate sample thickness determination would lead to the distortion of the reconstruction results. These distortions have the form of the periodic fluctuations appeared in the material parameter curves of the sample. To predict the appearance of these mistakes and to reduce the required accuracy of thickness determination one could use the procedure of numerical thickness correction during the THz spectrometer signal processing. According to this procedure the problem of functional minimization (3) needs to be solved for all possible sample thicknesses from the interval $[l_0 - \Delta l;\ l_0 + \Delta l]$, where l_0 is the thickness measured with the micrometer or the Vernier compass, and Δl is a possible mistake of the sample thickness measurement. The result of this solution is functions $n_1(f, l)$ and $\alpha_1(f, l)$ (the material parameter curve dependence on the sample thickness). The criteria of the curve smoothness:

$$R(l) = \left[\int_{f_{min}}^{f_{max}} \left[\zeta \left(\frac{\partial n_1(f, l)}{\partial f} \right)^2 + \varsigma \left(\frac{\partial \alpha_1(f, l)}{\partial f} \right)^2 \right] df, \right]^{-1} \tag{6}$$

is used to choose the smoothest pair of material parameter curves, where ζ and ς are the coefficients used to regulate the impact of the functions $n_1(f, l)$ and $\alpha_1(f, l)$ on the $R(l)$, f_{min} and f_{max} define the range of integration and correspond to the reliable frequencies for complex refractive index determination This sample thickness clarification approach helps to significantly improve the accuracy of reconstruction especially if the relation between ζ and ς is defined by:

$$\zeta = 10\varsigma.$$

Figure 3 illustrates several minimum branches of an error function. One of the presented minima of the error functional corresponds to the desired sample material parameters. Other minima (side branches) appear in correspondence to the multiple of 2π phase shifts of the time-domain signal Fourier-harmonics. In consequence, in the figure 4 the optical properties of the test samples are presented. This dependence of optical properties on frequency was obtained as the result of the global minimization of the sample pulse function in a wide range of the sample material parameter values.

Note, that additional conditions were used during the minimization procedure to prevent choosing of the false functional minimum. These conditions are based on the criteria of the optical material parameter curves smoothness. The points marked on the material parameter curves (figure 5) illustrate the results of the same test samples studying with THz backward wave oscillator spectrometer [10],[11]. One could see that difference between the results of the same sample studying with two different types of THz spectroscopy is lower than 1%.

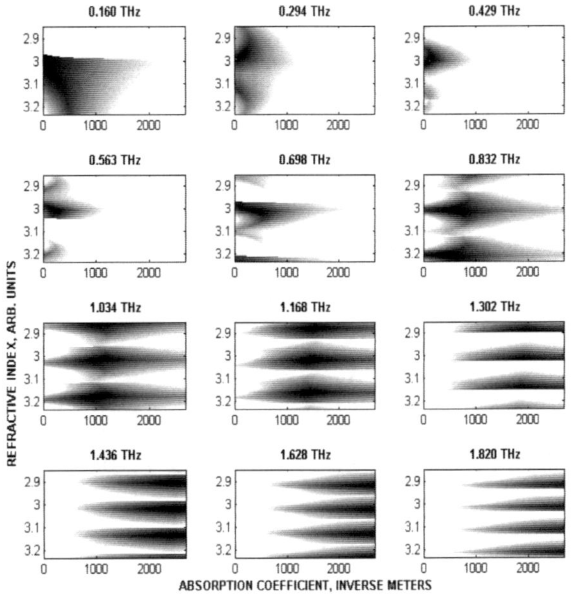

Figure 3. Error functional corresponding to the *ceramic* sample.

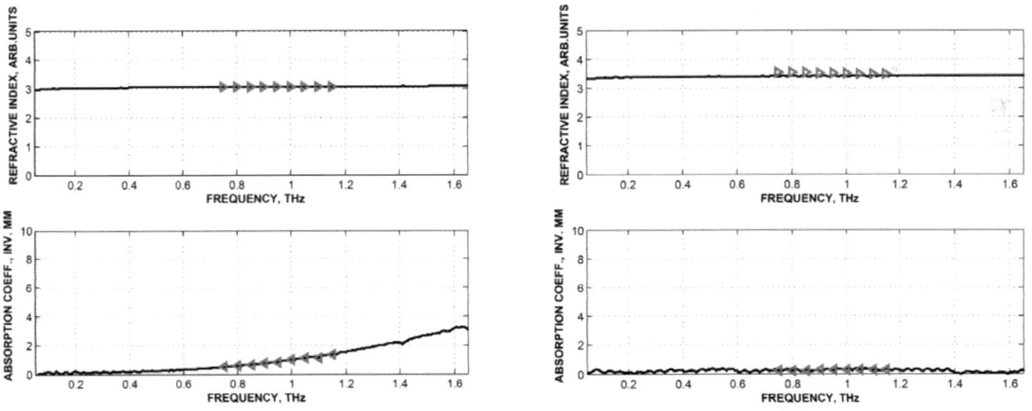

Figure 4. Material parameters of the test samples.

Conclusions

The new algorithm for material parameter curves determination was presented and implemented. *GaAs* and *ceramic* samples were studied experimentally. The procedure for sample thickness numerical clarification was offered. It significantly improves the accuracy of the material parameter curves reconstruction and reduces the required accuracy of the sample thickness *priori* measurements. The material parameter curves of two samples were reconstructed and compared to the results of the same sample studying utilizing THz backward wave oscillator spectroscopy. The presented algorithm could provide reconstruction error smaller than 1%.

Acknowledgments
This work was partially supported by grants from the Ministry of Education and Science of Russian Federation (№ 14.B37.21.0898, № 14.B37.21.1282) and from the Russian Foundation for Basic Research (RFBR 12-08-31104; RFBR 12-08-33112, RFBR 14-08-00175, RFBR 14-08-31124).

References
[1] Auston D H 1975 Picosecond Optoelectronic Switching and Gating in Silicon *Appl. Phys. Lett.* **26** 101 – 103
[2] Grischkowsky D, Keiding S, Vanexter M 1990 Far-Infrared Time-Domain Spectroscopy with Terahertz Beams of Dielectrics and Semiconductors *J. Opt. Soc. Am. B* **7** 2006 – 2015
[3] Duvillaret L, Garet F, Coutaz J-L 1999 Highly precise determination of Optical Constants and sample thickness in Terahertz Time-Domain Spectroscopy *Appl. Opt.* **38** 409 – 415
[4] Dorney T, Baraniuk R, Mittleman D 2001 Material parameter estimation with Terahertz Time-Domain Spectroscopy *J. Opt. Soc. Am. A* **18** 1562 – 1571
[5] Pupeza I, Wilk R, Koch M 2007 Highly accurate optical material parameter determination with THz time-domain spectroscopy *Opt. Express.* **15**(7) 4335 – 4350
[6] 9. Zaytsev K I, Karasik V E, Fokina I N, Alekhnovich V I 2013 Invariant embedding technique for medium permittivity profile reconstruction using terahertz time-domain spectroscopy *Opt. Eng.* **52**(6) 068203-1 – 068203-10
[7] Zaytsev K I, Fokina I N, Fedorov A K, Yurchenko S O 2012 Water and ice spectral response analysis in the terahertz spectral region at phase transition process *Herald of Bauman Moscow State Technical University. Series: "Natural sciences"* **S6** 113 – 119
[8] Alekhnovich V I, Zaytsev K I, Karasik V E 2013 Medium dielectric permittivity profile reconstruction using terahertz time-domain spectroscopy *Herald of Bauman Moscow State Technical University. Series: "Natural sciences"* **2** 50-66
[9] Vodopyanov K L 2008 Optical THz-wave generation with periodically-inverted GaAs *Laser & Photonics Review* **2**(1–2) 11 – 25
[10] Kozlov G V, Volkov A A 1998 Coherent Source Submillimeter Wave Spectroscopy *Millimeter and Submillimeter Wave Spectroscopy of Solids. Topics in Applied Physics* **74** 51 – 109
[11] Gorshunov B, Volkov A, Spektor I, Prokhorov A, MukhinA, Dressel M, Uchida S, Loidl A 2005 Terahertz BWO-spectroscopy *Int. J. Infrared and Millimeter Waves* **26**(9) 1217 – 1240 doi: 10.1007/s10762-005-7600-y

Quantum behaviour of water molecule in gemstone: terahertz fingerprints

Elena S Zhukova[1,2,3], Boris P Gorshunov[1,2,3], Victor I Torgashev[4], Vladimir V Lebedev[2,5], Gil'man S Shakurov[6], Reinhard K Kremer[7], Efim V Pestrjakov[8], Victor G Thomas[8], Dimitry A. Fursenko[8] and Martin Dressel[2]

[1]A M Prokhorov General Physics Institute, Russian Academy of Sciences, Russia
[2]Physikalisches Institut, Universität Stuttgart, Germany
[3]Moscow Institute of Physics and Technology, Russia
[4]Faculty of Physics, Southern Federal University, Russia
[5]Landau Institute for Theoretical Physics, Russian Academy of Sciences, Russia
[6]Kazan Physical-Technical Institute, Russian Academy of Sciences, Russia
[7]Max-Planck-Institut für Festkörperforschung, Germany
[8]Institute of Laser Physics, Russian Academy of Sciences, Russia

E-mail: gorshunov@ran.gpi.ru

Abstract. We have shown that a weak interaction of a lone H2O molecule with the "walls" of nano-sized crystalline cage of gemstone (beryl) results in emergence of a rich set of molecular vibrational states. By analogy with translational and librational bands in liquid water and ice corresponding absorption bands are explained as due to translational (T) and librational (L) movements of the H2O molecule which is hydrogen bonded to the cage walls. In beryl crystal lattice, however, the six-fold symmetry of the cage brings about additional effect of splitting of the T and L bands into fine structure due to tunnelling within the six-well potential relief. The presented results will be of use for analysis of more complicated systems with confined water molecules like H2O chains in carbon nano-tubes, molecular clusters in e.g. zeolites, clays, silica gels and other natural or synthetic frameworks, as well as for interfacial water in biological systems.

1. Introduction and background

Water is a chemical compound most pervasive on the Earth, liquid is most studied by physics, chemistry and biology [1]. Nevertheless, despite a simple structure of the H_2O molecule, water probably is the least understood of all liquids because of the complex network of hydrogen bonds. Water in contact with solid surfaces [2],[3] or single macromolecules, for instance proteins [4], forms a hydration layer in which a gradual transition from bound towards bulk water takes place. Although the characteristic network becomes less dynamic, weak hydrogen bonds allow librations, i.e. restricted rotations of the water molecules, which survive even in solid ice. Here we investigate single water molecules confined to nano-scale cages present in the crystal lattice of gemstone beryl $Be_3Al_2Si_6O_{18}$. Such nano-cages provide a well-defined environment for the water molecules that can be modeled and understood in great detail.

Content from this work may be used under the terms of the Creative Commons Attribution 3.0 licence. Any further distribution of this work must maintain attribution to the author(s) and the title of the work, journal citation and DOI.
Published under licence by IOP Publishing Ltd

2. Results

Using coherent source terahertz spectroscopy in combination with infrared Fourier-transform spectroscopy, we have measured the water-related absorption spectra in beryl crystal for two principle polarizations (electrical field vector of radiation **E** parallel and perpendicular to the c-axis) at frequencies 7 cm^{-1} to 7000 cm^{-1} and at temperatures from 300 K down to 5 K. We have observed quantum behavior of the confined H_2O molecules of type-I and type-II (figure 1). Via tunneling processes the wave functions of the type-I molecules hybridize and the discrete energy levels broaden to form bands. In highly anisotropic absorption spectra we identify transitions between and within these bands related to librational and translational motions similar to those present in ice and liquid water.

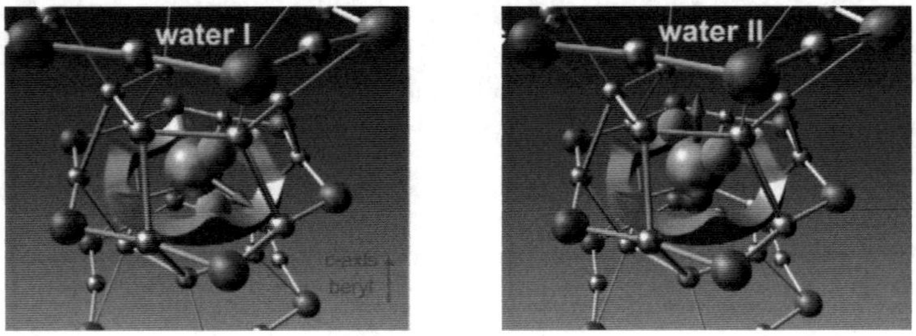

Figure 1. Honeycomb crystal structure (space group P6/mcc) of beryl contains open nano-channels that extend along the c-direction. They contain cavities of 5.1 Å diameter connected by bottlenecks of 2.8 Å size. Crystal water molecules enter the cavities and are oriented with the dipole moment (red arrows) either perpendicular (A) or parallel (B) to the c-direction. Hydrogen bonds weakly connect the water molecules to surrounding oxygen atoms as indicated by the green belt that resembles the periodic potential.

References

[1] Lynden-Bell R M, Morris S C, Barrow J D, Finney J L, Harper C L 2010 *Water and Life* (CRC Press, Boca Raton)
[2] Thiel P A and Madey T E 1987 The interaction of water with solid surfaces: fundamental aspects *Surf. Sci. Reports* **7** 211
[3] Henderson M A 2002 The interaction of water with solid surfaces: fundamental aspects revisited *Surf. Sci. Reports* **46** 1
[4] Ball P 2008 Water as an Active Constituent in Cell Biology *Chem. Rev.* **108** 74

Journal of Physics: Conference Series **486** (2014) 012020 doi:10.1088/1742-6596/486/1/012020

Parametric emission generation in cubic noncentrosymmetrical crystals

Gleb M Katyba and Vladimir S Gorelik

Bauman Moscow State Technical University, 2nd Baumanskaya str., 5, Moscow, 105005, Russia

E-mail: micalych@mail.ru

Abstract. The physics of polariton radiation generation in cube noncentrosymmetrical crystals is discussed. Theoretical values of polariton radiation frequencies in GaP, ZnSe, ZnTe and GaAs crystals were found. The basic circuit of parametrical terahertz radiation generator is introduced.

1. Introduction and background

The possibilities of spectroscopy in the terahertz (THz) frequency range have been increased with appearance of the novel techniques for generation and detection of THz waves. Spectroscopy of the polaritonic Raman scattering is the known method for studying of the collective excitations (phonons, magnons and others) in crystals [1]. The researching of the polaritonic Raman scattering spectrum has been performed for different kinds of the cubic noncentrosymmetrical crystals by using of different lasers. These technique allows the full characterization of the complex dielectric constant for many types of dielectric crystals [2]-[4]. Moreover, studying of the polaritonic Raman scattering is an important problem for development of novel coherent infrared radiation sources with tunable frequency (for example, THz- radiation source) [3],[5].

In the present paper the conditions of the polaritonic generation in cubic noncentrosymmetrical crystal for collinear geometry of Raman scattering were discussed.

2. Results

We are using the dispersion equation $\omega(k)$ for a monochromatic polariton wave, propagating in a dielectric crystal [3]:

$$\omega^2 = \frac{c_0^2 k^2}{\varepsilon(\omega)\mu(\omega)} \tag{1}$$

where $\varepsilon(\omega)$ is a dielectric permittivity , $\mu(\omega)$ is a magnetic permeability (in the subsequent calculation we introduced $\mu(\omega)=1$), k is a wave vector, ω is a frequency. The dispersion of the dielectric permittivity is assumed as the well-known Kurosawa ratio [1] for the isotropic dielectric medium with the one resonance:

Content from this work may be used under the terms of the Creative Commons Attribution 3.0 licence. Any further distribution of this work must maintain attribution to the author(s) and the title of the work, journal citation and DOI.
Published under licence by IOP Publishing Ltd

$$\varepsilon(\omega) = \frac{\varepsilon_{\infty}\left(\omega_L^2 - \omega^2\right)}{\left(\omega_T^2 - \omega^2\right)} \tag{2}$$

where the constant $\varepsilon_{\infty} = n_{\infty}^2$ is a dielectric permittivity of the medium corresponding to the high frequency range, ω_L, ω_T are zeroes of the frequency and the pole of the dielectric function. Taking into account the laws of conservation of momentum and energy in the elementary process of the Raman scattering by polaritons the following equation could be written [3]:

$$\begin{cases} \omega = \omega_0 - \omega_1 \\ k = k_0 - k_1 \end{cases}. \tag{3}$$

Here ω, ω_0 , ω_1 are the frequencies of polariton, exciting and scattered radiation, respectively, and k, k_0, k_1 are appropriate wave vectors. Note, the equation (3) corresponds to the collinear scattering geometry. An expression for the polariton frequency could be obtained after elementary conversions:

$$\nu = \sqrt{\frac{\left(n_0 + \nu_0\left(\frac{\partial n}{\partial \nu}\right)_{\nu=\nu_0}\right)^2 \nu_T^2 - \varepsilon_{\infty}\nu_L^2}{\left(n_0 + \nu_0\left(\frac{\partial n}{\partial \nu}\right)_{\nu=\nu_0}\right)^2 - \varepsilon_{\infty}}} \tag{4}$$

where n_0 is a refractive index of the crystal corresponding to the exciting radiation with frequency ν_0, $\nu = \omega/2\pi c_0$ is a frequency in cm^{-1}. Using [6]-[8] we could find the frequency of THz radiation generation in the cube noncentrosymmetrical crystals corresponding to the collinear Raman scattering geometry. These frequencies are listed in the table 1.

Table 1. Polariton frequency for different cube nonsentrosimmmetrical crystals at collinear geometry of the Raman scattering: the first column shows the crystal type, the second column shows the pump frequency, the third column shows the refractive index at the pump frequency, and the fourth column gives the frequency of generated radiation.

Type of crystall	ν_0, cm^{-1}	n_0	ν, cm^{-1}
GaP	15803	3.315	302,4
ZnSe	18797	2.6797	139
ZnTe	15803	2.9625	171,7
GaAs	9434	3.4851	175

For instance in the figure 1 the polariton curves for GaP crystal are presented. Point **A** corresponds to the frequency of the THz radiation generation in the case of the collinear geometry of polariton light scattering.

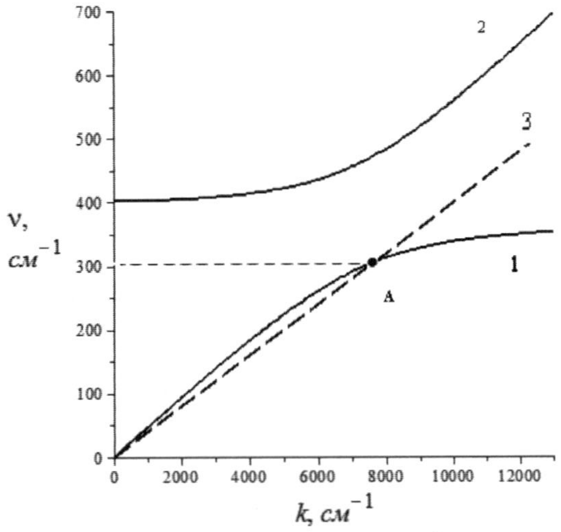

Figure 1. Dispersion curves of the GaP polaritons (1,2) and linear curve $\nu(k) = |k_0 - k_1|/2\pi$ (3). The frequency of the exciting radiation was set to $\nu_0 = 15803\,см^{-1}$.

Monochromatic THz-radiation generation in considered cubic noncentrosymmetrical crystals may be carried out according to the scheme presented in figure 2.

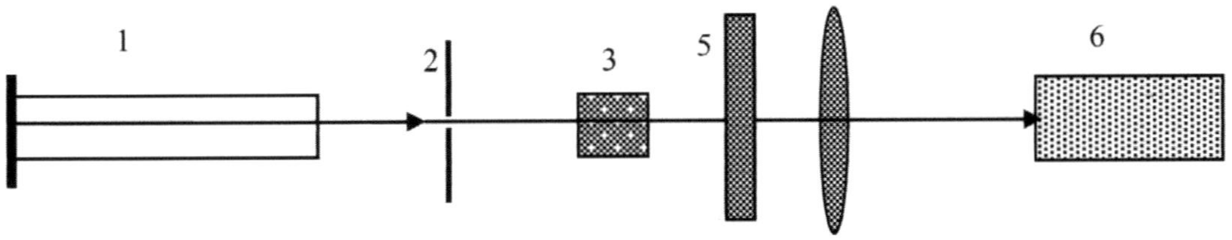

Figure 2. An experimental set-up for studying the process of the parametric THz radiation generation: 1 is an optical laser, 2 is a diaphragm, 3 is a GaP crystal, 4 is a silicon lens, 5 is a silicon flat, 6 is a THz radiation receiver.

The silicon flat 5 presented in the figure 2 is used both as a filter cutting the part of the optical pumping light which could partially transmit through the crystal sample 3 and as an additional Fabry-Perot resonator enhancing the efficiency of THz lasing on a certain frequencies.

Thus, polariton frequencies (or parametric radiation frequencies) were calculated for GaP, ZnSe, ZnTe and GaAs crystals in the cases of pumping by ultra short optical pulses of different types of lasers with high average and peak power. Conventional scheme of THz radiation generator based on these results was introduced.

References

[1] Litvinov O S, Gorelik V S 2006 *Electromagnetic Waves and Optic* (Bauman Moscow State Technical University, Moscow, Russia)

[2] Born M, Huang K 1954 *Dynamical Theory of Crystal Lattices (Oxford Classic Texts in the Physical Sciences)* (Clarendon Press, Oxford, UK)

[3] Polivanov Yu N 1978 Raman scattering of light by polaritons *Sov. Phys. Usp.* **21**(10) 805–831

[4] Hebling J, Almasi G, Kozma I Z, Kulh J 2002 Velosity matching by pulse front tilting for large area THz-pulse generation *Opt. Express* **10** 1161-1166
[5] Hirori H, Doi A, Blanchard F, Tanaka K 2011 Single-cycle terahertz pulses with amplitudes exceeding 1 MV/cm generated by optical rectification in $LiNbO_3$ *Appl. Phys. Lett.* **98**(9) 091106-1 – 091106-1
[6] Blistanov A A, Bondarenko V S, Perelomova N V et al 1982 *Acoustic Crystals*, ed M P Shaskolskaya (Nauka, Major Edition Physical and Mathematical Literature, Moscow, Russia)
[7] Babichev A P, Babushkina N A, Bratkovsky A M et al 1991 *Physical Quantities: Reference Book*, ed Grigorieva I S and Meylihova E Z. (Energoatomizdat, Moscow, Russia)
[8] Sonin A S and Vasilevska A S 1971 *Electro-Optical Crystals* (Atomizdat, Moscow, Russia)

Ultrasensitive superconducting terahertz detectors: novel approaches and emerging materials

Andrei Sergeev[1], Vladimir Mitin[1], Boris Karasik[2] and Sergey Vitkalov[3]

[1]University at Buffalo, the State University of New York, USA
[2]Jet Propulsion Laboratory, California Institute of Technology, USA
[3]City College of the City University of New York, New York 10031, USA

E-mail: asergeev@buffalo.edu

Abstract. Novel approaches to THz sensing based superconductor detectors and emerging superconducting nanomaterials have a strong potential to boost development of advanced optoelectronic devices, such as THz detectors, THz mixers, single photon counters and quantum calorimeters with outstanding sensitivity. Such devices have a number of applications in THZ environmental and industrial monitoring, astrophysics, homeland security, and medicine. Single photon counters have potential as key elements for optical communication and networking, quantum imaging and metrology, quantum optical computing and bio-photonics, and single-molecule spectroscopy.

1. Introduction and background

The bolometer was invented 125 years ago by S.P.Langley. In preliminaries to his original publication he pointed out: "I found … that science possessed no instrument that could deal successfully with quantity of radiant heat so minute… I have entered into these preliminary remarks as an explanation of the necessity of such an instrument as that which I have called the Bolometer or Actinic Balance." Originally, bolometers were designed for measuring incident electromagnetic radiation. However, these devices are sensitive not only to the electromagnetic but to every form of energy. For example, bolometers may be used to measure phonon fluxes and have a potential to measure separate phonons. Bolometric devices are used for search for unknown forms of energy, like dark matter. Currently, superconducting bolometers are the most sensitive detectors in the THz range.

Superconducting bolometer was suggested just before the World War II. These devices combine small heat capacity with large intrinsic responsivity, which is determined by the strong temperature dependence of resistance at the superconducting transition. Currently, superconducting microbolometers are widely used in submillimeter astrophysics, high-energy physics, quantum calorimetry, etc. Various applications of bolometers require a wide range of device operating times. Moreover, the sensitivity and operating time are interrelated. Improving one of this characteristics results in deterioration of another. For these reasons, it is extremely desirable to have possibility to adjust the operating time to values required by specific applications. However, in "classical" bolometers on solid substrates the thermal coupling between a superconductor (thermometer) and substrate (heat sink) is determined by the Kapitza conductance or, in other words, by the phonon transparencies at the film-substrate interface. At low temperatures, the phonon transparency is given

Content from this work may be used under the terms of the Creative Commons Attribution 3.0 licence. Any further distribution of this work must maintain attribution to the author(s) and the title of the work, journal citation and DOI.
Published under licence by IOP Publishing Ltd

Journal of Physics: Conference Series **486** (2014) 012021 doi:10.1088/1742-6596/486/1/012021

by the acoustic mismatch between film and substrate, and this parameter cannot be verified substantially.

Proposed three decades ago [1], superconducting hot-electron bolometers (HEBs) combine robustness and reliability with small electron heat capacity and controllable thermal conductance. In HEBs, radiation overheats only electrons, while film phonons play the role of a heat sink for electrons. This regime is achieved by decreasing the film thickness, so that the phonon escape from the film to substrate dominates over the phonon-electron relaxation in the film.

2. Superconducting nanohebs

We investigate capabilities of a a submicron size superconducting transition edge sensor made from thin disordered *Ti* film, which has a critical temperature, $T_C \approx 350$ mK [2]. The *Ti* bridge is fabricated on silicon or sapphire substrate between *Nb* contacts which block the diffusion of hot electrons out of the bridge due to Andreev reflection (see figure 1). A planar antenna provides coupling to THz radiation. The *Ti* nanobridge will operate in the voltage-biased TES mode; that is, its operating temperature T will be somewhat lower than the critical temperature, T_c, and the resistance at the operating point, R, will be much smaller than the normal resistance R_n. This situation is convenient for the simultaneous antenna impedance match and the SQUID noise match. Indeed, the dc resistance in the operating point is < 1 Ohm that is the device Johnson noise is greater than the noise of a typical dc SQUID. At the same time, the radiation "sees" the device normal resistance that is 40-50 Ohm, which can be well matched using planar antennas. The radiation causes an increase of the electron temperature, which relaxes due to the electron-phonon interaction with the time constant $\tau_{e\text{-}ph}$. Depending on the bias conditions, the response time can decrease, that is, $\tau = \tau_{e\text{-}ph}/(1+L)$, where $L \sim 10\text{-}100$ is the electro-thermal feedback (ETF) loop gain. The expected Noise Equivalent Power, $NEP = T(2k_B G_{e\text{-}ph})^{1/2}$, based on the experimentally measured electron-phonon thermal conductance $G_{e\text{-}ph} = C_e/\tau_{e\text{-}ph}$ and on the dimensions of available *Ti* nanosensors is shown in Fig. 2. The electron-phonon time is greatly enhanced in disordered thin films. This circumstance helps to achieve high sensitivity. One can see that in this *photon integration mode* the device intrinsic *NEP* reaches 10^{-19} W/Hz$^{1/2}$ [2].

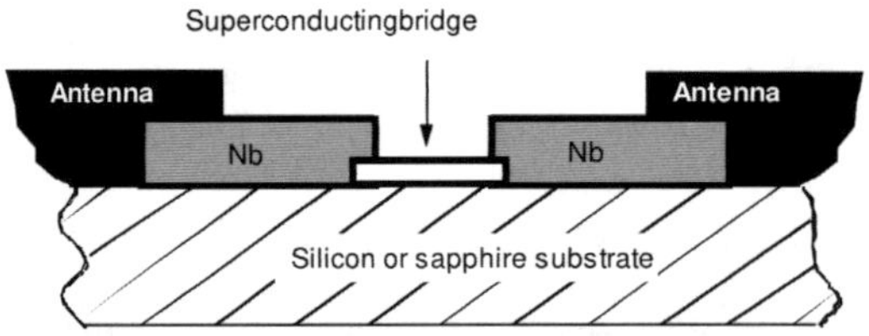

Figure 1. Cross-sectional view of the HEDD design. The energy gap in the HEDD is suppressed due to operation at the superconducting transition edge.

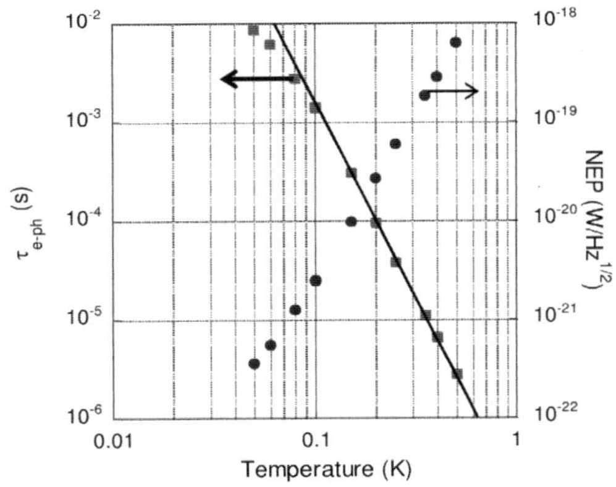

Figure 2. The experimentally measured electron-phonon relaxation time in disordered *Ti* film and the expected *NEP* for a submicron device.

3. Superconducting counters of terahertz photons

The *photon counting mode* for THz spectroscopy has been considered before rather as an option than the actual practical requirement for spectrometer on SAFIR-like platforms [3],[4]. It is becoming clear that it will be crucial for achieving the highest sensitivity needed above 1 THz. The ability of the bolometric device to detect a single photon depends on the magnitude of the intrinsic thermal fluctuations. The equivalent energy fluctuation is $\delta E = 2T((2n)^{1/2}C_e/\alpha)^{1/2}$, where $\alpha = d\ln R/d\ln T$ and n = 5-6 is the exponent in the electron-phonon thermal conductance, which depends on the purity of material. In order to detect low energy photons, the temperature must be low and the volume of the device must be small. For the photon counter, the frequency $\nu_R = \delta E/h$ can be treated as the "red boundary" or lower frequency limit for the detection mechanism. Our estimates show that $\nu_R < 1$ THz can be achieved for a *Ti* device of practical submicron size at 0.3 K. In other words, the equivalent *NEP* is 10^{-20} W/Hz$^{1/2}$ at 0.3 K. The best integrating detector alternatives with comparable sensitivity would require cooling down to at least 0.1 K.

The value of $N_{max.} = 1/(2\tau)$ determining the maximum count rate, is about 5×10^5 sec^{-1}. This would provide a large dynamic range for the counter. In order to achieve the background limited performance, the dark count rate in the device itself, N_d, must be lower than photon arrival rate. The fundamental dark counts are due to the fact that the device phonon noise produces spikes which can be mistaken for the signal. To avoid that, a corresponding level of the discrimination threshold, $E_T < h\nu$, must be chosen in the photon counter or/and in the readout electronics. Then the counts are caused only by the energy fluctuations that exceed the discriminator threshold. The corresponding dark count rate is given by

$$N_d = \frac{1/(2\tau)}{\sqrt{2\pi}} \cdot \int_{E_T/\delta E}^{\infty} \exp\left(-x^2/2\right)\, dx$$

With the above parameters, setting $E_T = 0.7h\nu$ brings N_d to ~ 0.1 count per sec. At the same time, the probability to detect a real photon (= quantum efficiency) remains high. N_d is very sensitive to the ratio ET/dE. If, for example, the device length is chosen to be 0.4 μm instead of 0.5 μm, Nd becomes less than 0.01.

The upper limit for the spectral range of the counter depends on the superconducting transition

Journal of Physics: Conference Series **486** (2014) 012021 doi:10.1088/1742-6596/486/1/012021

width, δT_c. For example, for $\delta T_c \approx 15$ mK, a 2-THz photon would drive the counter into the normal state. Photons of higher energy will still be detectable, but the amplitude of the output pulses will be truncated at the level corresponding 2 THz.

A dc SQUID can be the readout amplifier for the HEDD. Even with a feedback loop, the SQUID bandwidth can be easily made ~ 100 kHz, which is very close to the HEDD bandwidth. An array SQUID with the bandwidth ~ 1 MHz is even better choice. A typical SQUID has a spectral noise density ~ 1 pA/Hz$^{1/2}$ so the total current noise is a fraction of a nA. This is much smaller than the expected amplitude of the response.

4. Emerging superconducting materials for THz sensing

Impressive recent progress [5],[6] in the MBE growth of superconducting heterojunctions of *LaCuO/LaSrCuO* oxides provides new intriguing possibilities for novel superconducting hot-electron sensors. The transition temperature in these low-dimensional superconductors is considerably higher than in bulk crystals of $La_{2-x}Sr_xCuO_4$, reaching as high as 51K. Remarkably, superconductivity may persist even in a single atomic layer at the atomically sharp interface between two non-superconducting materials, namely, the Mott insulator La_2CuO_4 and the strongly over-doped metallic oxide $La_{2-x}Sr_xCuO_4$.

New low-dimensional superconducting materials have the following physical characteristics that make them especially attractive for use in detectors, THz heterodyne receivers, and IR single photon counters: (i) High superconducting transition temperature, adjustable by doping in a wide range from sub-Kelvin to 50 K; (ii) Ultra-small electron heat capacity, C_e. due to the quasi-two-dimensional nature of the electron gas in oxide heterostructures; (iii) Strong coupling to the radiation, expected due to the Drude absorption (short momentum relaxation time); (iv) Strong photoresponse due to strong temperature dependence of the resistivity at the transition; (v) Strong electron heating by THz radiation, due to strong absorption of the radiation, small electron heat capacity, and significant temperature dependence of the resistivity; (vi) Ideal phonon heat sink for hot electrons, because nonequilbrium phonons immediately escape from the superconducting layer (there is no Kapitza resistance for phonons); (vii) Weak electron coupling to thermal phonons (see below). New materials provide exciting possibilities for superconducting hot electron nanobolometers that may lead to potential breakthroughs in modern THz sensing technologies.

To evaluate performance of hot electron bolometers based on superconducting MBE-grown quasi-two dimensional LaSrCuO structures, we investigate electric and thermal characteristics of the resistive states in structures with several (3-10) superconducting Cu-O layers covered by top and bottom overdoped metallic layers. The resistive states in these S-M sandwiches are observed over a wide temperature range (8-25 K), which shifts to lower temperatures in a magnetic field. At low currents the *V-I* characteristic demonstrate a strong nonlinearity, $V = \gamma I^2$, where the coefficient γ is mainly determined by the resistance, $R(T,H)$, and weakly depends on T and H directly.

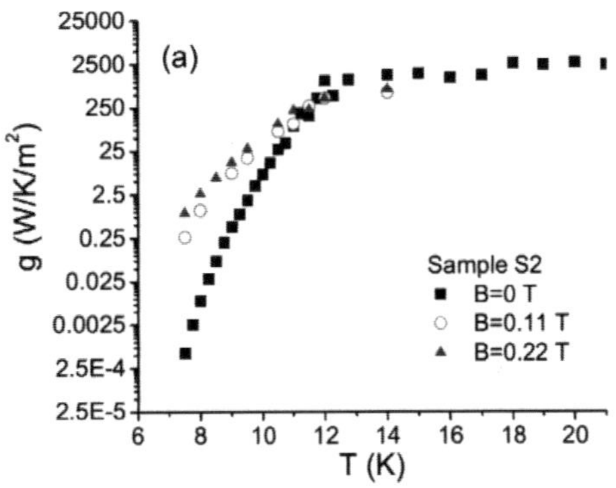

Figure 3. Electron-phonon thermal conductance of superconducting LaSrCuO nanostructure calculated from the nonlinearity of V-J characteristic.

Figure 3 presents the temperature dependence of the electron-phonon thermal conductance per unit area at different magnetic fields. Let us note that the resistance of the superconducting layer is shunted by top and bottom metallic sheets. In result, a characteristic temperature T*, at which the resistance of the superconducting layer reaches a half of its normal resistance, corresponds to a significant total resistance of the sandwich. For the sample S2 the value of T* is approximately 13K. Figure3 demonstrates that above T* the thermal conductance slowly increases with the temperature indicating a small temperature variation of the heat capacity C_e and electron-phonon relaxation time $\tau_{e\text{-}ph}$, which is consistent with the heating model. Below T* the electron-phonon conductance strongly decreases with the temperature decrease.

Above T* the conductance is expected to be close to the conductance in the normal state. For the sample S2 the conductance is ~2000 W/Km^2 at 14 K. Other samples show approximately the same values of the thermal conductance.

Let us first highlight that these values are three orders in magnitude smaller than that in NbN or MgB ultrathin films. Such strong difference may be explained by the small carrier density in LaSrCuO, which leads to a small electron heat capacity and weak electron-phonon coupling via the deformation potential. Let us highlight that in the metallic state the electron-phonon relaxation rate is proportional to the carrier density [7]. Weak coupling of hot electrons to the thermal phonons strongly increases the responsivity and sensitivity of HEBs based on MBE-grown LaSrCuO structures.

In summary, the detection of single THz photons is vital for many branches of basic science, as well as for a variety of engineering applications. Sensors of these quanta are of great interest for molecular spectroscopy, nanoscale solid-state physic and thermophysics. Research in this area is of fundamental importance for the broad field of mesoscopic solid-state physics and, in particular, for numerous quantum nanodevices operating at low temperatures. Superconducting single photon THz detectors are expected to be a valuable tool for NASA, European, and Russian missions addressed to investigations of structure and evolution of the Universe.

References

[1] Gershenzon E M, Gershenzon M E, Goltsman G N, Semenov A D and Sergeev A V 1984 Heating of quasi-particles in a superconducting film in the resistive state *JETP Lett.* **34** 268-270

[2] Wei J, Olaya D, Karasik B S, Pereverzev S V, Sergeev A V, Gershenson M E 2008 Ultrasensitive hot-electron nanobolometers for terahertz astrophysics□*Nat. Nanotechnology* **3** 496-500

[3] Karasik B S, Sergeev A and Prober D 2011 Nanobolometers for THz photon detection *Inaugural Issue of IEEE Trans. THzz Sci. Technol.* **1** 97

[4] Karasik B and Sergeev A 2005 Hot-electron superconducting photon counter *IEEE Appl.*

Supercond. **15** 618-621

[5] Gozar A, Logvenov G, Kourkoutis L F, Bollinger A T, Giannuzz L A, Muller D A, Bozovic I 2008 High-temperature interface superconductivity between metallic and insulating cuprates *Nature* **455** 782

[6] Bollinger A T, Dubuis G, Yoon J, Pavuna D, Misewich J and Bozovic I 2011 Superconductor-insulator transition in LaSrCuO at the pair quantum resistance *Nature* **472** 458-460

[7] Sergeev A and Mitin V 2000 Electron-phonon interaction in disordered conductors: Static and vibrating scattering potentials *Phys. Rev. B.* **61** 6041

Journal of Physics: Conference Series **486** (2014) 012022 doi:10.1088/1742-6596/486/1/012022

Theory of plasmonic terahertz detection by a dual-grating-gate field-effect transistor

Denis V Fateev, Vyacheslav V Popov and Sergey A Nikitov

Kotelnikov Institute of Radio Engineering and Electronics of RAS and Saratov State University, Russia

E-mail: FateevDV@yandex.ru

Abstract. Plasmonic terahertz detection in a dual-grating-gate field-effect transistor structure with an asymmetric unit cell is studied theoretically and modeled numerically. It is shown that DC bias current flowing in the structure channel strongly affects the detector responsivity.

1. Introduction and background

Grating-gate field-effect transistor (FET) structures are promising for plasmonic detection of terahertz (THz) radiation [1]. For symmetry reasons, the rectification of THz radiation can reveal itself only if an asymmetry is introduced into the grating-gate FET structure. The asymmetry can be introduced in different ways: either by applying DC drain bias current in the grating-gate-FET channel (electrical asymmetry) or by designing a periodic grating gate with an asymmetric unit cell (geometrical asymmetry). It was shown that the plasmonic THz detector based on a dual-grating-gate (DGG) FET structure with an asymmetric unit cell (A-DGG-FET detector), see figure 1, demonstrates strong photovoltaic response (up to 10 kV/W) without passing DC electric current through the device channel [2]. In this paper, we study the effect of DC bias current passing through A-DGG-FET channel on the detector responsivity.

2. Results

We simulated plasmonic THz detection in InAlAs/InGaAs/InP A-DGG-FET structure with the electrophysical parameters typical for room temperature by using the perturbative approach described in [3],[4] for $w_1 = 200$ nm, $w_2 = 1600$ nm, $s_1 = 200$ nm, and $s_2 = 400$ nm (see figure 1). Different gate voltages U_{G1} and U_{G2} can be applied to different interdigitated sub-gratings having the metal finger widths w_1 and w_2, respectively. A DC drain bias current j_0 passes through the A-DGG-FET channel. The photocurrent j_{ph} induced in the device channel leads to an additional voltage drop between the source and drain terminals, which in effect is the photovoltage $U_{ph} = j_{ph}R$, where R is the device resistance.

Content from this work may be used under the terms of the Creative Commons Attribution 3.0 licence. Any further distribution of this work must maintain attribution to the author(s) and the title of the work, journal citation and DOI.
Published under licence by IOP Publishing Ltd

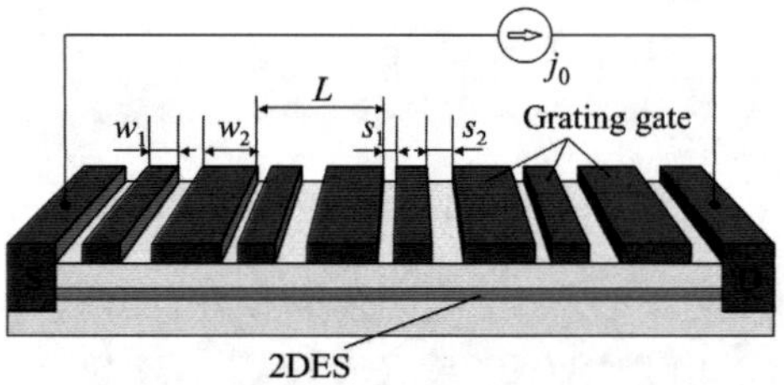

Figure 1. Schematic view of the A-DGG-FET THz detector.

Figure 2 shows the normalized THz photovoltage for different values of the DC drain bias current for U_{G1} = 0 and U_{G2} = −1.6 V. Two sources of asymmetry compete in the presence of DC drain current: (i) a geometric asymmetry of the structure unit cell and (ii) electrical asymmetry caused by the DC drain current. The effect of the DC bias current is much more pronounced at low frequencies so that the THz photovoltage changes its sign with changing the polarity of the DC current in A-DGG-FET channel. However, the geometrical asymmetry dominates at high frequencies so that DC current changes only a value of the photoresponse leaving the sign of the photoresponse unaffected.

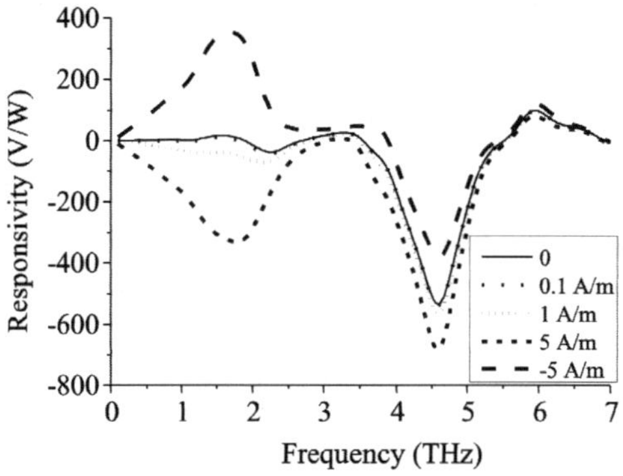

Figure 2. Terahertz photovoltage between the drain and source contacts of the A-DGG-FET detector with asymmetry factor $s_1/s_2 = 0.5$ for different values of DC bias current.

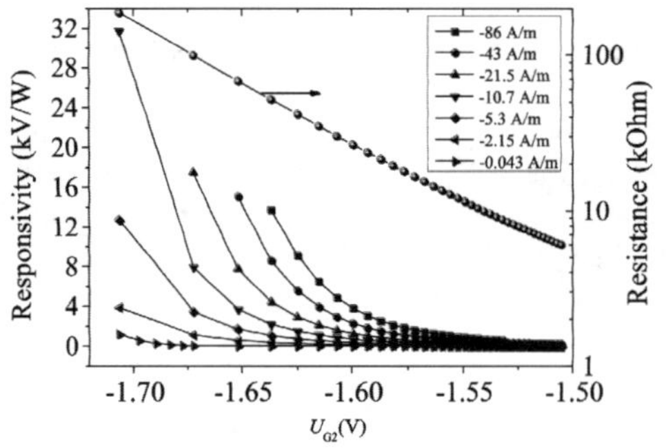

Figure 3. Responsivity vs the gate voltage U_{G2} for $U_{G1} = 0$ for different values of DC bias current (left axis) at frequency 1.5 THz and the channel resistance (right axis).

Acknowledgements
This work has been supported by RFBR (Grant No. 12-02-00813) and by the Russian Academy of Sciences Program “Fundamentals of Nanostructure Technologies and Nanomaterials”. The work was performed under umbrella of the GDR-I project "Semiconductor Sources and Detectors for Terahertz Frequencies".

References

[1] Popov V V 2011 Plasmon excitation and plasmonic detection of terahertz radiation in the grating-gate field-effect-transistor structures *J. Infrared Millim. Terahertz Waves* **32**(10) 1178-1191

[2] Popov V V, Fateev D V, Otsuji T et al 2011 Plasmonic terahertz detection by a double-grating-gate field-effect transistor structure with an asymmetric unit cell *Appl. Phys. Lett.* **99**(24) 243504-1 – 243504-4

[3] Aizin G R, Popov V V, Polischuk O V 2006 Plasmon enhanced electron drag and terahertz photoconductance in a grating-gated field-effect transistor with two-dimensional electron channel *Appl. Phys. Lett.* **89**(14) 143512-1–143512-3

[4] Aizin G R, Fateev D V, Tsymbalov G M, Popov V V 2007 Terahertz plasmon photoresponse in a density modulated two-dimensional electron channel of a GaAs/AlGaAs field-effect transistor *Appl. Phys. Lett.* **91**(16) 163507-1–163507-3

Journal of Physics: Conference Series **486** (2014) 012023 doi:10.1088/1742-6596/486/1/012023

Terahertz and infrared surface plasmon-polaritons in double-graphene layer structures

Dmitry Svintsov[1], Vladimir Vyurkov[1], Victor Ryzhii[2,3] and Taiichi Otsuji[3]

[1] Institute of Physics and Technology, Russian Academy of Science, Russia
[2] Research and Educational Center "Photonics and Infrared Technology", Bauman Moscow State Technical University, Russia
[3] Research Institute of Electrical Communication, Tohoku University, Japan

E-mail: svintcov.da@mipt.ru

Abstract. Spectra and damping of surface plasmon-polaritons (SPPs) in double graphene layer structures are studied theoretically. The SPPs in those structures exhibit a remarkable voltage tunability of velocity and damping, inherent to gated graphene. The spatial dispersion of conductivity significantly augments the free path and cutoff frequency of SPPs. It is also pointed out to the effect of photon-assisted interlayer tunneling, which leads to a substantial decrease in SPP damping.

1. Device model and main equations

The structures consisting of two graphene layers separated by a thin dielectric (figure 1) look promising for practical applications in nano- and optoelectronics. A bias voltage applied between graphene layers (GLs) controls over carrier densities in them, which boosts a functionality of double-graphene optical modulators [1]. The bias can also operate on interlayer tunneling in transistor-like structures [2].

Here we study the propagation of plasma waves in double graphene layers. The spectra and damping are obtained from a general dispersion relation for surface plasmon-polaritons

$$\frac{\varepsilon}{\sqrt{\mathbf{k}^2 - \varepsilon\omega^2 / c^2}} = -\frac{4\pi i}{\omega}\sigma_{\mathbf{k},\omega}$$

where ω is a SPP frequency and $\mathbf{k}$ is a SPP wave vector, ε is a permittivity of environment, c is a speed of light, and $\sigma_{\mathbf{k},\omega}$ is graphene conductivity. The latter depends both on frequency and wave vector and incorporates interband and intraband transitions. At small separation between GLs the interlayer tunneling also contributes to conductivity; the latter contribution is estimated via Fermi golden rule and second-order perturbation theory.

2. Results

It is shown that the bias voltage shifts the Fermi energies ε_F in GLs thus changing the interband conductivity and the propagation length of SPPs. Particularly, SPPs with frequencies $\omega > 2\varepsilon_F / \hbar$ are strongly damped due to interband electron transitions [3]. SPPs with lower frequencies exhibit propagation lengths up to 1 μm if the graphene conductivity is determined by electron-phonon scattering (figure 2).

Content from this work may be used under the terms of the Creative Commons Attribution 3.0 licence. Any further distribution of this work must maintain attribution to the author(s) and the title of the work, journal citation and DOI.
Published under licence by IOP Publishing Ltd

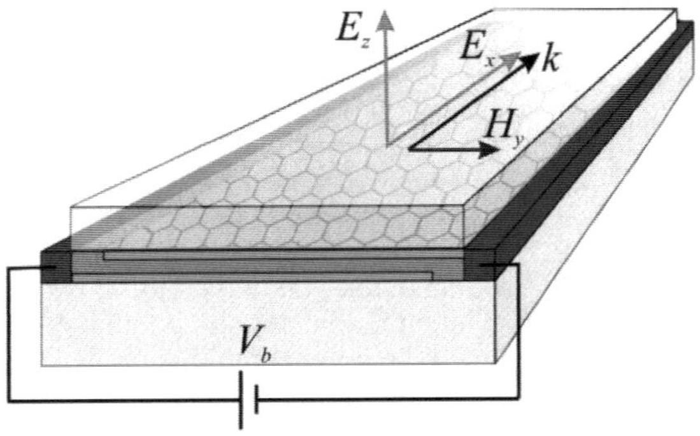

Figure 1. Schematic view of double-graphene layer structure with directions of wave-vector k, electric E and magnetic H fields in SPP.

In present calculations we rigorously treat the spatial dispersion of graphene conductivity omitted in the previous studies of plasmons in graphene. It turned out to be crucial and leads to much longer SPP free paths (up to 100 times) compared to that in preceding evaluations.

Finally, we emphasize that in double-graphene layer structures with steady bias between GLs the interlayer tunneling accompanied by photon emission is feasible. In the case the Fermi energies in layers differ resulting in a population inversion which can lead to negative dynamic tunnel conductivity. The latter effect causes an increase in SPP propagation length.

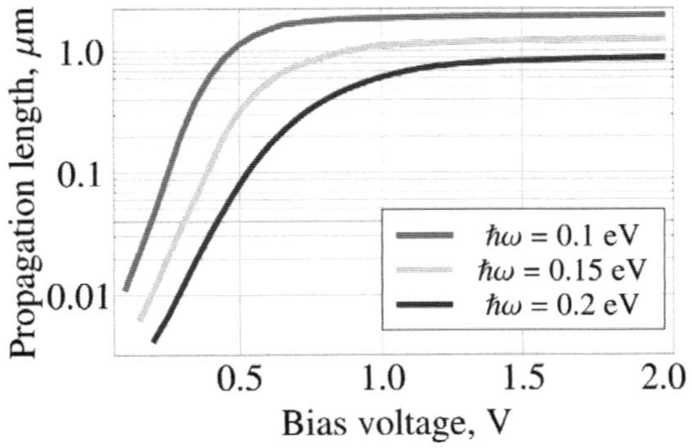

Figure 2. SPP propagation length in double GL structure vs. bias voltage.

References

[1] Liu M, Yin X and Zhang X 2012 *Nano Lett.* **12**(3) 1482–1485
[2] Britnell L et al 2012 *Science* **335** 947
[3] Svintsov D et al 2013 *J. Appl. Phys.* **113** 053701

Sensing of phase transition in medium with terahertz pulsed spectroscopy

Kirill I Zaytsev, Irina N Fokina, Aleksey K Fedorov and Stanislav O Yurchenko

Bauman Moscow State Technical University, 2nd Baumanskaya str., 5, Moscow, 105005, Russia

E-mail: kirzay@gmail.com

Abstract. Phase state identification and phase transition registration in condensed matter are significant applications of terahertz spectroscopy. A set of fundamental and applied problems are associated with the phase state problem. Our report is devoted to the experimental analysis of the spectral characteristics of water and water solution during the phase transition from the solid state to the liquid state via the method of terahertz pulsed spectroscopy. In this work transformation of the sample spectral characteristics during the phase transition were observed and discussed. Possible application of terahertz pulsed spectroscopy as an effective instrument for phase transition sensing was considered.

1. Introduction and background

There are many fundamental and applied problems associated with the study of medium phase transitions which could be solved by the use of terahertz (THz) pulsed spectroscopy. For example, one of them is an important problem of technological non-destructive evaluation during a process of vacuum molding of the polymer composite materials. There are a lot of polymer vacuum molding technologies [1],[2]. All of them involve fabrication of a polymer composite using polymer phase transition under the change of any internal or external environment conditions, for example: changes of the medium chemical compound, changes of the polymer temperature, etc. In order to reduce the wastes of materials in the vacuum molding technologies, several tasks of the polymer molding process control need to be solved. It includes the polymer medium homogeneity control, control of the package filling completeness, control of the phase transition process in polymer medium during the composite manufacturing, etc. Note, during the vacuum molding process, polymer is located in a vacuum package which is transparent for THz rays and opaque for visual and infrared radiation, thus visual and infrared light could not be used for solving control task. This paper shows that the THz pulsed spectroscopy system could sense the phase transitions in media, and since the molding package is transparent for the THz radiation, all of the listed control problems could be successfully solved with THz pulsed spectroscopy.

The region between 0.1 and 10.0 THz is commonly referred to THz region [3]. THz rays are non-ionizing in nature, and are able to penetrate a certain depth into a large number of dielectric materials. THz spectral absorption lines of media are associated with the presence of molecular low-energy rotational and vibrational degrees of freedom. Absorption is caused by changes in energy levels of molecule movements induced by THz rays. The more complex the medium molecule is the more complex THz absorption spectrum of molecule becomes. Also the more polar molecule is the more powerful THz absorption line is.

Content from this work may be used under the terms of the Creative Commons Attribution 3.0 licence. Any further distribution of this work must maintain attribution to the author(s) and the title of the work, journal citation and DOI.
Published under licence by IOP Publishing Ltd

Because of all these properties there are a lot of potential applications of THz technologies, including spectroscopy of biological tissue [4],[5], medical diagnostics [6]-[9], detection of concealed weapons, explosives and drugs [10],[11], nondestructive evaluation of constructional materials [12], art inspection [13], gas sensing [14],[15] etc. Since absorption of THz light in various media is caused by the changes in molecule energy levels associated with different molecule rotational and vibrational states, and the spectrum of possible molecule states transforms significantly during the medium phase transition, THz spectral characteristics of medium also undergo great transformations during the phase transition. This fact provides the ability to use THz spectra for studying and controlling processes of phase transitions in media. This property of THz spectra is beneficial in different fundamental and applied aspects of THz science. And it could be used for technological non-destructive evaluation of constructional materials, in particular for control the vacuum molding process, mentioned above, during manufacturing of polymer composite materials.

This work shows the ability for studying the processes of phase transition in a medium via the instrument of THz pulsed spectroscopy using the ice-water transition as an example.

Recently much attention is paid to the investigation of chemical, physical and biological properties of water [17]-[21] and aqueous solution in THz range of the electromagnetic spectrum. Water molecule is strongly polar and has many rotational and vibrational degrees of freedom, thus the THz absorption spectrum of water is very complex. In gas state water molecule has symmetric and asymmetric stretches and bending of the covalent bonds. In liquid state the motion and the structure of water are more complex because of the long-range interaction between molecules due to the hydrogen bonding [22]. Note, the absorption of water is much higher in THz range in contrast to microwave and optical ranges of the electromagnetic spectrum [23]. Paper [24] is dedicated to determination of the liquid water complex dielectric permittivity model in the THz range. It was shown, that the THz complex dielectric permittivity of water could be accurately described with the double Debye model. Moreover, information about microscopic mechanisms, leading to water anomalous properties, is contained in the temperature dependence of its THz absorption spectrum. Ronne et al. [25] determined the optical constants of liquid water as a function of temperature in the discrete temperature scale between 271.1 and $366.7\ K$ and in the frequency range from 0.1 to $2.0\ THz$.

2. Results

This work shows the results of the “ice-water” phase transition studying via the method of THz pulsed spectroscopy. The THz reflectivity transformation during the samples of pure water and 5% water sugar solution defrosting are presented and discussed.

The main idea of the terahertz pulsed spectroscopy (TPS) can be described as follows. Short pulses of THz radiation with a wideband spectrum (from $0.1\ THz$ to at least $3.5\ THz$) are utilized to measure THz optical properties of media. The electrical field of the THz pulse $E_{THz}(t)$ is detected with high time resolution up to $50.0\ fs$ after transmission or reflection of the pulse from sample surface. It is possible to analyze complex amplitude transmission or reflection coefficients of a sample in a wide frequency range via the procedure of the fast Fourier transformation, applied to the detected time-domain signal.

For our purposes we have used a TPS system fully described in paper [26]. It was complemented by the thermostat module as it is shown in the figure 1. TPS system utilizes ultra-short optical pulses for both generation of THz radiation in a $LT - GaAs$ photoconductive antenna and detection of THz pulses in a $ZnTe$ electrooptical detector. The used TPS system is able to detect THz spectral characteristics of a sample in a frequency range from 0.1 to $3.0\ THz$ in reflection and transmission modes with a repetition rate up to $400.0\ Hz$. Such a high repetition rate could be achieved due to the usage of the ultra-fast delay stage in the TPS system. Averaging of 10 signals was applied during the measurements to provide satisfactory signal-to-noise ratio (SNR) and dynamic range [27] (DNR) in the TPS time domain data. Therefore the sample spectrum characteristics were registered with the frequency rate $40.0\ Hz$ during the experiment.

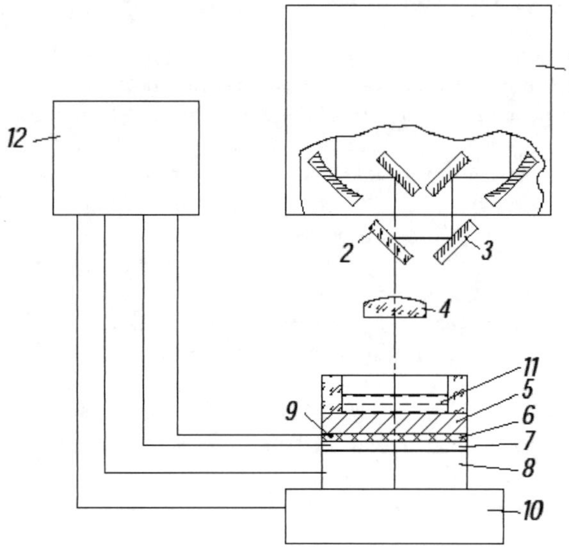

Figure 1. Schematic representation of the sample holder with thermostat: 1–4 are TDS spectrometer; 5 is a metal plate; 6 is a layer of thermopaste; 7 is a Peltier element; 8,10 are parts of the thermocouple; 9 is a sample temperature sensor; 11 is a sample cuvette of water and aqueous solutions; 12 is an electronic control unit.

Two signals need to be obtained for reconstruction of the sample THz reflectivity: $E_r(t)$ is the signal reflected from the reference surface with a very high and homogeneous reflectivity in a wide spectral range, and $E_s(t)$ is the sample signal reflected from the sample surface. The reference signal needs to be detected once (before the experiment), and the sample signal needs to be detected on each step of the sample temperature variation. A planar gold mirror was used to obtain the reference signal. As the signals were detected, the sample reflectivity reconstruction could be accomplished according to the following equation:

$$\tilde{R}(\nu_t) = \frac{\tilde{E}_s(\nu_t)}{\tilde{E}_r(\nu_t)}, \tag{1}$$

where $\tilde{E}_r(\nu_t)$ and $\tilde{E}_s(\nu_t)$ are the Fourier spectra of the reference and the sample signals, respectively. Note, equation (1) should be applied to every sample signal obtained on the each step of the sample temperature variation; therefore the temporal dependence of the sample reflectivity during the phase transition would be obtained.

The sample of water and the sample of 5% water solution of sugar were placed in a cuvette as it can be seen from figure 1. The samples were studied with TPS system in the reflection mode during the samples defrosting. Since THz beam was focused on the air-medium interface with a high aperture lens, the THz spot diameter close to $1.22 \cdot \lambda$ was achieved. Temperature of the samples was controlled with a thermostat built of thermocouple and Peltier element. The sample was connected with the Peltier element via the metal plate and the thermopaste. The temperature of the samples varied continuously within the interval between -10 and $+20$ °C and then in steps of 5°C with the accuracy of the temperature keeping equal to ± 0.1 °C. The samples reflectivity was measured both at the different stationary samples temperatures and during the continuous changes of the temperature.

The recorded time dependences of the THz amplitude reflectivity variations during the phase transition in the media are presented in the figure 2a (water sample) and figure 2b (5% sugar aqueous solution sample). In order to notice the clear differences and rather high contrast between solid and liquid states reflectivity one could consider figure 3, which contains the results of samples reflectivity determination at the discrete stationary temperatures: -5, $+5$, $+10$, $+15$ and $+20$ °C.

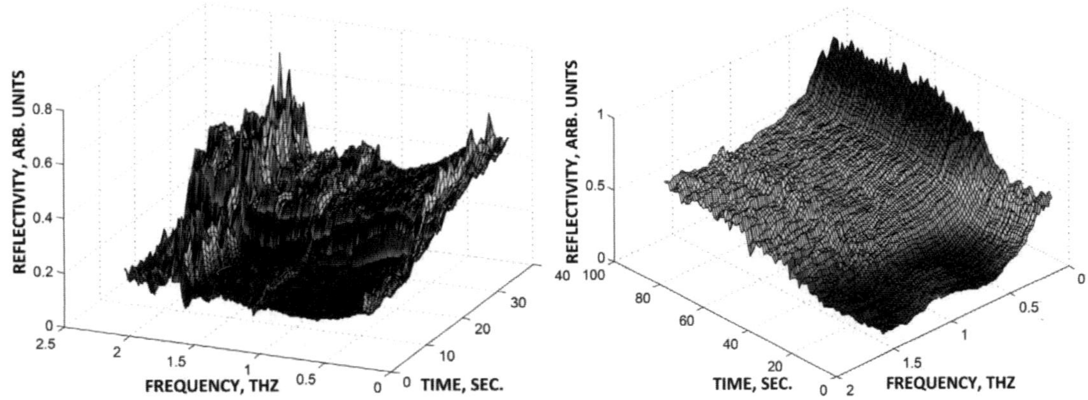

(a) (b)

Figure 2. Time dependence of the sample reflectivity modulus during the phase transition in the medium: (a) corresponds to the water sample, (b) corresponds to the water sugar solution sample.

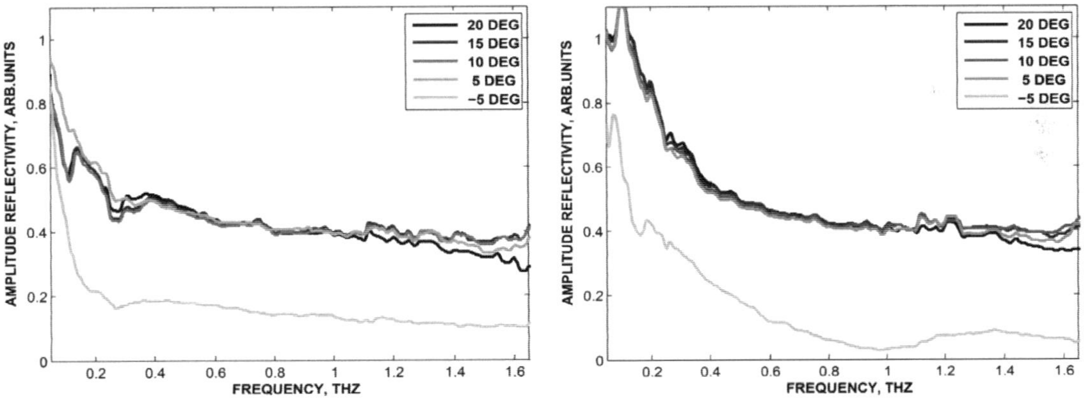

(a) (b)

Figure 3. Amplitude reflectivity modulus at the different discrete sample temperatures: (a) corresponds to the water sample, (b) corresponds to the 5% water sugar solution sample.

In complete accordance with the theoretical conception [28],[29], spectral characteristics of the samples were transformed during the phase transition. Therefore, in both our experiments phase transition was registered. There are several strong lines in the spectral reflectivity curves (figures 2,3), especially at the frequencies above 1.6 THz. These lines are caused by the fluctuations of the water vapor content along the THz beam path during our experiment. To prevent the appearance of this lines in future works one should vent the THz beam path with nitrogen. Reflectivity of the second sample is higher than 1.0 $arb.units$ at low frequencies. This is due to the low signal-to-noise ratio of the TDS system in defined frequency range.

The TDS system utilized in the present work could be used for detection of medium phase transitions with the temporal resolution higher than $0.1\ sec$. The temporal resolution of the considered method is limited both by the repetition rate of the THz spectrum registration, and by the size of the spot of THz beam focused on the surface of the sample. Phase transition of the medium does not occur simultaneously at every point of the medium-air interface. The transition starts from several active points of the sample surface and runs as a wave in the medium-air interface. Since this wave runs through the THz beam spot for a finite time, the THz pulsed spectrometer obtained the average THz reflectivity of the liquid and the solid state regions of the sample surface which are on the THz beam path. The velocity of the phase transition wave depends on a large numbers of factors, including the degree of the sample overcooling, but it is constant for the fixed experiment conditions. Possible ways of raising the temporal resolution of the phase transition sensing is reduction of the THz spot size and raising of the TDS system repetition rate. Modern THz pulsed spectrometers could reach $500.0\ Hz$ repetition rate of the spectrum registration. Being equipped with a high aperture optical system focusing the THz beam on the sample surface, modern TDS system could provide temporal resolution higher than $0.01\ s$.

Conclusions

The ability of the THz pulsed spectroscopy usage for studying of the medium phase transitions have been discussed. The results of the water sample and the water solution of sugar sample studying during the medium phase transition were obtained. The dependence of the samples reflectivity versus the experiment time was shown. High contrast between the reflectivity of the samples in the liquid and the solid states was observed. The ability of THz pulsed spectroscopy usage for non-destructive sensing of the polymer phase transition during the process of the polymer composite material vacuum molding was suggested.

Acknowledgments

This work was partially supported by grants of the Ministry of Education and Science of Russian Federation (№ 14.B37.21.0898, № 14.B37.21.1282) and of the Russian Foundation for Basic Research (RFBR 12-08-31104; RFBR 12-08-33112). A.K.F. is an RQC Fellow.

References

[1] van Rijswijk K, Bersee H E N, Beukers A, Picken S J, van Geenen A A 2006 Optimisation of anionic polyamide-6 for vacuum infusion of thermoplastic composites: Influence of polymerisation temperature on matrix properties *Polymer Testing* **25**(3) 392-404

[2] Goren A, Atas C 2008 Manufacturing of polymer matrix composites using vacuum assisted resin infusion molding *Archives of Materials Science and Engineering* **34**(2) 117-120

[3] Lee Y-S 2009 *Principles of Terahertz Science and Technology* (Springer, New York, NY)

[4] Nazarov M M, Shkurinov A P, Kuleshov E A, Tuchin V V 2008 Terahertz time-domain spectroscopy of biological tissues *Quantum Electronics* **38**(7) 647-654

[5] Pickwell E, Cole B E, Fitzgerald A J, Wallace V P, Pepper M 2004 Simulation of terahertz pulse propagation in biological systems *Appl. Phys. Lett.* **84**(12) 2190 – 2192

[6] Wallace V P, Woodward R M, Fitzgerald A J, Pickwell E, Pye R J, Arnone D D 2003 Terahertz-pulsed imaging of cancers *Proc. SPIE* **4949** 353 – 359

[7] Zaytsev K I, Karasik V E, Koroleva S A, Fokina I N 2012 Study the ability of terahertz imaging systems application for medical diagnosis *Herald of Bauman Moscow State Technical University. Series: "Instrumentation"* **4** 114 – 127

[8] Fitzgerald A J, Wallace V P, Jimenez-Linan M, Bobrow L, Pye R J, Purushotham A D, Arnone D D 2006 Terahertz Pulsed Imaging of Human Breast Tumors *Radiology* **239**(2) 533–540

[9] Yu C, Fan S, Sun Y, Pickwell-MacPherson E 2012 The potential of terahertz imaging for cancer diagnosis: A review of investigations to date *Quantum Imaging in Medicine and Surgery*

2(1) 33 – 45

[10] Murrill S R, Franck C C, Espinola R L, Petkie D T, De Lucia F C, Jacobs E L 2011 Enhanced terahertz imaging system performance analysis and design tool for concealed weapon identification *Proc. SPIE* **8188** 81880J-1 – 81880J-15

[11] Federici J F, Schulkin B, Huang F, Gary D, Barat R, Oliveira F, Zimdars D 2005 THz imaging and sensing for security applications – explosives, weapons and drugs *Semicond.Sci. Technol.* **20**(7) S266 – S280

[12] Stoik C D, Bohn M J, Blackshire J L 2008 Nondestructive evaluation of aircraft composites using transmissive terahertz time domain spectroscopy *Opt. Express* **16**(21) 17039 – 17051

[13] Abraham E, Younus A, Delagnes J C, Mounaix P 2010 Non-invasive investigation of art paintings by terahertz imaging *Appl. Phys. A* **100**(3) 585 – 590

[14] Jacobsen R H, Mittleman D M, Nuss M C 1996 Chemical recognition of gases and gas mixtures with terahertz waves *Opt. Lett.* **21**(24) 2011 – 2013

[15] Mittleman D M, Jacobsen R H, Neelamani R, Baraniuk R G, Nuss M C 1998 Gas Sensing Using Terahertz Time-domain Spectroscopy *Appl. Phys. B* **67**(3) 379 – 390

[16] Mittleman D M, Nuss M C, Colvin V L 1997 Terahertz spectroscopy of water in inverse micelles *Chem. Phys. Lett.* **275**(3-4) 332 – 338

[17] Globus T, Bykhovski A, Khromova T, Gelmont B, Tamm L K, Salay L C 2007 Low-terahertz spectroscopy of liquid water *Proc.SPIE* **6772** 67720S-1 – 67720S-11.

[18] Bunkin N F, Yurchenko S O, Suyazov N V, Shkirin A V 2012 Structure of the nanobubble clusters of dissolved air in liquid media *J. Biolog. Phys.* **38**(1) 121 – 152

[19] Gorshunov B P, Zhukova E S, Torgashev V I, Lebedev V V, Shakurov G S, Kremer R K, Pestrjakov E V, Thomas V G, Fursenko D A, Dressel B 2013 Quantum Behavior of Water Molecules Confined to Nanocavities in Gemstones *J. Phys. Chem. Lett.* **4**(12) 2015 – 2020

[20] Aliev I N, Yurchenko S O 2010 Evolution of perturbations of a charged interface between immiscible inviscid fluids in the interelectrode gap *Fluid Dynamics* **45**(5) 817-826

[21] Bunkin N F, Yurchenko S O, Suyazov N V, Shkirin A V 2012 Structure of the nanobubble clusters of dissolved air in liquid media *J. Biol. Phys.* **38**(1) 121-152

[22] Siegel P H 2004 THz Technology in Biology and Medicine *IEEE Trans. Microwave Theory Tech.* **52**(10) 2438 – 2448

[23] Xu J, Plaxco K W, Allen S J 2006 Absorption spectra of liquid water and aqueous buffers between 0.3 and 3.72 THz *J. Chem. Phys.* **124**(3) 036101-1 – 036101-3

[24] Pickwell E, Cole B E, Fitzgerald A J, Pepper M, Wallace V P 2004 Simulation of terahertz pulse propagation in biological systems *Appl. Phys. Lett.* **84**(12) 2190 – 2192

[25] Ronne C, Thrane L, Astrand P O, Wallqvist A, Mikkelsen K V, Keiding S R 1997 Investigation of the temperature dependence of dielectric relaxation in liquid water by THz reflection spectroscopy and molecular dynamics simulation *J. Chem. Phys.* **107**(14) 5319–5331

[26] Zaytsev K I, Karasik V E, Fokina I N, Aleckhnovich V I 2013 Invariant embedding technique for medium permittivity profile reconstruction using terahertz time-domain spectroscopy *Opt. Eng.* **52**(6) 068203-1 – 068203-10

[27] Naftaly M, Dudley R 2009 Methodologies for determining the dynamic ranges and signal-to-noise ratios of terahertz time-domain spectrometers *Opt. Lett.* **34**(8) 1213 – 1215

[28] Lee D C, Chapman D 1986 Infrared spectroscopic studies of biomembranes and model membranes *Bioscience Reports* **6**(3) 235 – 256

[29] Lima Jr J A, Freire P T C, Melo F E A, Filho J M, Fischer J, Havenith R W A, Broer R, Bordallo H N 2013 Using Raman spectroscopy to understand the origin of the phase transition observed in the crystalline sulfur based amino acid l-methionine *Vibrational Spectroscopy* **65** 132 – 141

Journal of Physics: Conference Series **486** (2014) 012025 doi:10.1088/1742-6596/486/1/012025

AlGaN/GaN plasmonic terahertz electronic devices

Michael Shur

ECSE and PAPA, Rensselaer Polytechnic Institute, Troy, NY 12180, USA

E-mail: shurm@rpi.edu

Abstract. A very large electron sheet density and a relatively long momentum relaxation time of the two-dimensional electron gas in III-N heterostructures makes this materials system to be very attractive for plasmonic electronics applications.

1. Introduction

Terahertz (THz) plasmonic devices [1],[2], are getting close to being commercialized. Efficient room temperature detectors implemented in AlGaAs/GaAs, AlGaN/GaN heterostructures [3] and Si CMOS terahertz cameras [4] have been reported. Graphene transistors exhibited overdamped plasmonic response, and graphene plasmonic detectors are expected to achieve superior performance. While most of the device efforts focused on THz plasmonic detectors, terahertz radiation emitted by short channel InGaAs [5] and AlGaN/GaN [6],[7] devices at cryogenic and room temperatures has been reported. Using "plasmonic crystals" [8], which are one-dimensional [9], two-dimensional or three-dimensional plasmonic arrays [10], could boost performance by orders of magnitude. The vast majority of the results deal with overdamped response related to nonlinearities in the device response. But the resonant plasmonic detection has also been reported [11],[12]. Recent results on extremely high sheet two-dimensional electron gas (2DEG) densities in AlInN/GaN structures [13], predictions of extremely high overshoot electron velocities in InN [14], and resonant terahertz response in AlGaN/GaN grating gate structures [15],[16] make this materials system very attractive for further development of plasma wave electronic devices. In this paper, we discuss how transport properties of short-channel III-N plasmonic devices and their performance.

2. Collision dominated and ballistic electron transport in III-N 2DEG

Figure 1 shows the measured [17] and calculated temperature dependence of 2DEG mobility, μ, in AlGaN/GaN heterostructure. (The mobility calculations accounted for optical phonon, acoustic phonon, and ionized impurity scattering.) Also shown is the ballistic mobility [18],[19] (i.e. the electron mobility limited by the device contacts given (for the degenerate 2DEG) by [20]

$$\mu_{bal} = \frac{2}{\pi} \frac{qL}{m_n v_F}$$

Here q is the electronic charge, L is the gate length, m_n is the electron effective mass, $v_F = \frac{\hbar}{m_n}\sqrt{2\pi n_s}$ is the Fermi velocity, and n_s is the electron sheet density. As seen, the overall effective mobility

Content from this work may be used under the terms of the Creative Commons Attribution 3.0 licence. Any further distribution of this work must maintain attribution to the author(s) and the title of the work, journal citation and DOI.
Published under licence by IOP Publishing Ltd

$$\mu_{eff} \approx \frac{\mu_{bal}\mu_{eff}}{\mu_{bal} + \mu_{eff}}$$

at temperatures below 250 K.

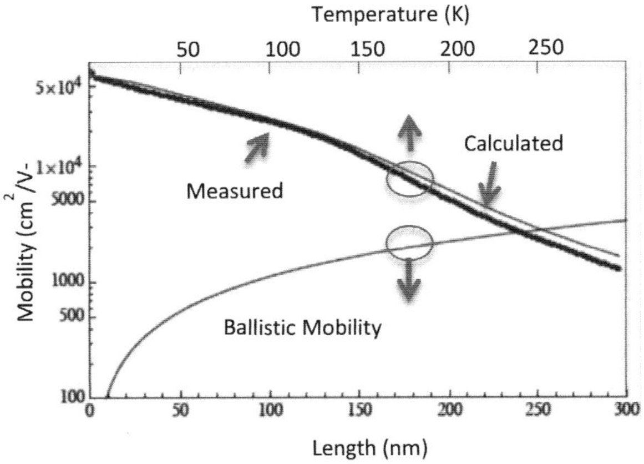

Figure 1. Measured and calculated temperature dependence of two-dimensional electrons in AlGaN/GaN heterostructure and ballistic mobility versus gate length.

Figure 2 shows the electron plasma $S = \sqrt{\frac{q^2 n_s d}{\varepsilon_0 \varepsilon}}$ and Fermi velocities as functions of the 2DEG sheet density in GaN.

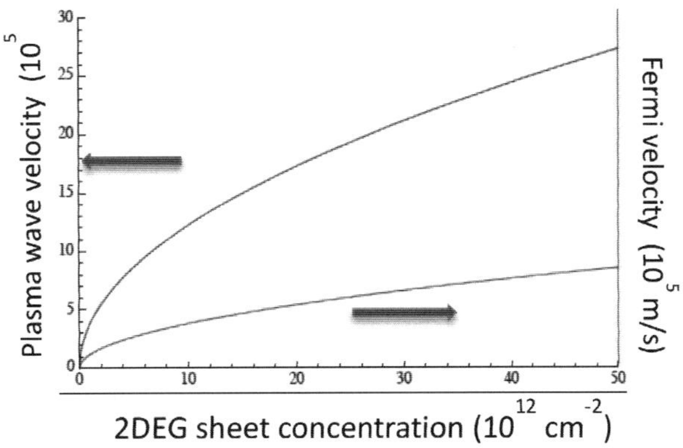

Figure 2. Plasma and Fermi velocities versus 2DEG sheet density.

As was shown in [21], the ratio of the decrement related to the Landau damping to the real part of the plasma frequency is given by

$$\frac{Im\,\omega}{Re\,\omega} = \frac{\sqrt{\pi}}{2}\frac{E_S^{3/2}}{E_F(k_B T)^{1/2}}\exp\left[\frac{E_F - E_S}{k_B T}\right]$$

As seen, the plasma wave velocity is much larger than the Fermi velocity in the III-N 2DEG Fermi velocity and, therefore, the Landau damping is relatively small.

Figure 3 shows the fundamental plasma frequency

$$f = \frac{S}{4L}$$

for III-N 2DEG FETs for short circuit boundary condition at the source side of the gated channel and open circuit condition at the drain side.

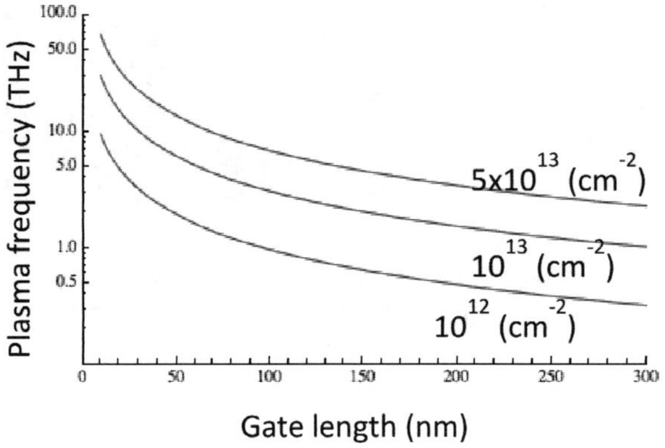

Figure 3. Fundamental plasma frequency versus gate length for different 2DEG densities.

Figure 4 shows the plasma wave quality factor calculated using the momentum relaxation time extracted from the measured mobility data shown in Figure 1 for frequencies of 1 THz and 5 THz.

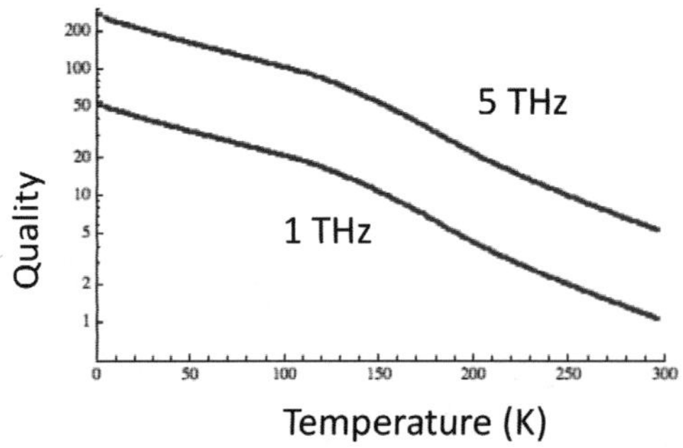

Figure 4. Plasma wave quality factor versus temperature for 1 THz and 5 THz plasma frequencies for III-N FETs.

These results show that the resonant detection and THz emission are possible at room temperature using short channel devices with the gate length below 50 to 70 nm and operating at frequencies of the order of 5 THz or above. Longer devices are expected to operate as broadband THz plasmonic detectors or in the intermediate regime with the quality factor on the order of 1.

Journal of Physics: Conference Series **486** (2014) 012025 doi:10.1088/1742-6596/486/1/012025

In the broadband detection regime, THz radiation couples to the device at the gate edges and excites an overdamped plasma mode that propagates into the channel along the distributive RC transmission line formed by the channel resistance and the gate-to-channel capacitance (see Figure 5).

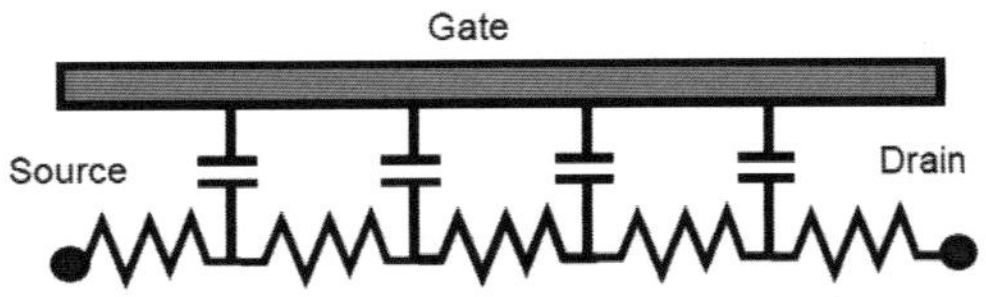

Figure 5. Lossy transmission line representation of FET channel for overdamped plasma wave propagation.

The characteristic scale of the decay depends on the field effect mobility μ and signal frequency ω (see figure 6).

$$L_0 = \sqrt{\frac{\mu V_{GT}}{\omega}}$$

Here V_{GT} is the gate voltage swing.

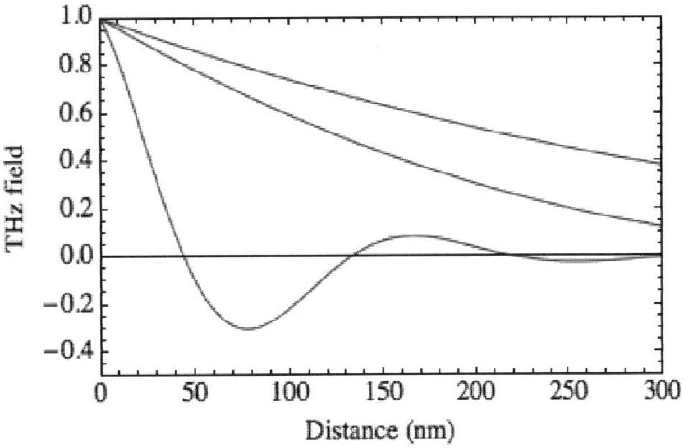

Figure 6. Plasmonic decay. Top, middle, and bottom curves are for f =0.2 THz, 0.5 THz, and 5 THz for GaN 2DEG.

Kachorovski et al. [22] calculated the conversion efficiency of a GaN-based FET defined as the ratio of the power dissipated by radiation-induced dc current to the THz dissipated power. And predicted the maximal value of the conversion efficiency on the order of 10%.

3. Experimental data

El Fatimy et al. reported on both resonant and non-resonant plasmonic detection GaN HEMTs with 150 nm gates (see figure 7) [23]. They reported on the minimum Noise Equivalent power of $5x10^{-9}$ $W/Hz^{0.5}$.

Taginawa et al. [3] achieved the responsivity of 1100 V/W using a GaN plasmonic detectors matched to 1 THz radiation.

Muravjov et al. [24] observed strong plasmon resonances in the terahertz transmission spectra in the range between 1 and 5 THz of large-area slit-grating-gate AlGaN/GaN-based FET structures at temperatures from 10 to 170 K. The resonance frequencies corresponded to the excitation of plasmons with wave vectors equal to the reciprocal lattice vectors of the metal grating, which coupled plasmons and incident terahertz radiation. The resonances tunable by the applied gate voltage were observed up to 170 K. The results were in a good agreement with the theoretical predictions [25].

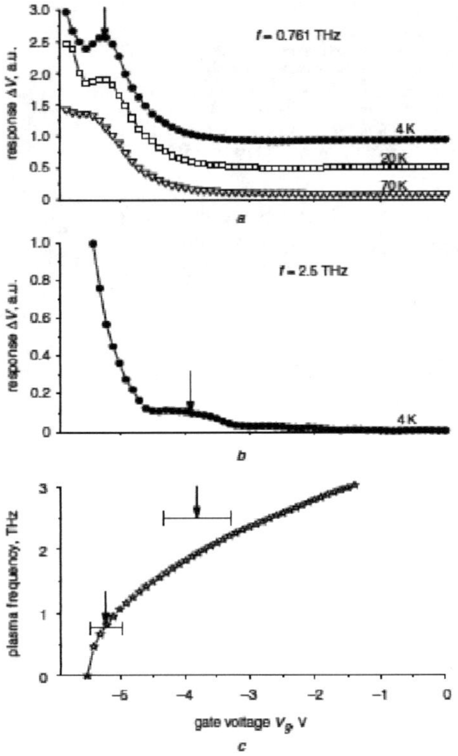

Figure 7. Plasmonic response of GaN/AlGaN -based FET to THz radiation. Bars show the position and width of resonant maxima. Line shows the calculated dependence of fundamental plasma frequency versus gate voltage (From [22]).

4. Conclusion

Exceptional transport properties of III-N 2DEG make GaN-based FETs uniquely suitable for applications in advanced plasmonic electronic devices for detection, mixing, and generation of THz radiation.

References

[1] Dyakonov M and Shur M S 1993 Shallow Water Analogy for a Ballistic Field Effect Transistor. New Mechanism of Plasma Wave Generation by DC Current *Phys. Rev. Lett.* **71**(15) 2465-2468

[2] Dyakonov M and Shur M 1996 Detection, mixing, and frequency multiplication of terahertz radiation by two-dimensional electronic fluid *IEEE Trans. ED* **43**(3) p. 380–387

[3] Tanigawa T, Onishi T, Takigawa Sh and Otsuji T 2010 Enhanced Responsivity in a Novel AlGaN / GaN Plasmon-Resonant Terahertz Detector Using Gate-Dipole Antenna with Parasitic Elements *DRC 2010, the University of Notre Dame, South Bend, IN, (June 21-23, 2010)* IV.A-9

[4] Boppel S, Lisauskas A, Max A, Krozer V and Roskos H G 2012 CMOS detector arrays in a virtual 10-kilopixel camera for coherent terahertz real-time imaging *Opt. Lett.* **37**(4) 536-538

[5] Knap W, Lusakowski J, Parenty T, Bollaert S, Cappy A, Popov V and Shur M S 2004 Emission

of terahertz radiation by plasma waves in 60 nm AlInAs/InGaAs high electron mobility transistors *Appl. Phys. Lett.* **84**(13) 2331-2333

[6] Deng Y, Kersting R, Xu J, Ascazubi R, Zhang X C, Shur M S, Gaska R, Simin G S, Khan M A and Ryzhii V 2004 Millimeter Wave Emission from GaN HEMT *Appl. Phys. Lett.* **84**(15) 70-72

[7] El Fatimy A, Dyakonova N, Meziani Y et al 2010 AlGaN/GaN high electron mobility transistors as a voltage-tunable room temperature terahertz sources *J. Appl. Phys.* **107** 024504

[8] Teperik T V, Garcia de Abajo F J, Popov V V and Shur M S 2007 Strong terahertz absorption bands in a scaled plasmonic crystal *Appl. Phys. Lett.* **90** 251910

[9] Kachorovskii V Yu and Shur M S 2012 Current-induced terahertz oscillations in plasmonic crystal *Appl. Phys. Lett.* **100** 232108

[10] Dmitriev A P, Kachorovskii V Yu and Shur M S 2007 Granular semiconductor/pyroelectric media as a tunable plasmonic crystal *Solid State Electron.* **51** 812–815

[11] Knap W, Deng Y, Rumyantsev S and Shur M S 2002 Resonant detection of subterahertz and terahertz radiation by plasma waves in submicron field effect transistors *Appl. Phys. Lett.* **81** 4637

[12] Otsuji T, Hanabe M and Ogawara O 2004 Terahertz plasma wave resonance of two-dimensional electrons in InGaP/InGaAs/GaAs high-electron-mobility transistors *Appl. Phys. Lett.* **85**(11) 219

[13] Sattu A, Billingsley D, Deng J, Yang J, Simin G, Shur M and Gaska R 2011 Low loss AlInN/GaN Microwave Switch *Electron. Lett.* **47**(15) 863–865

[14] O'Leary S K, Foutz B E, Shur M S and Eastman L F 2006 Steady-State and Transient Electron Transport Within the III-V nitride semiconductors, GaN, AlN, and InN: A Review *J. Mater. Sci.: Mater. Electron.* **17** 87–126

[15] Muravjov A V, Veksler D B, Popov V V, Polischuk O, Hu X, Gaska R, Pala N, Saxena H, Peale R E, Shur M S 2010 Temperature dependence of plasmonic terahertz absorption in grating-gate GaN HEMT structures *Appl. Phys. Lett* **96** 042105

[16] Popov V V, Pala N and Shur M S 2012 Room temperature terahertz plasmonic detection by antenna arrays of field-effect transistors *Nanoscience and Nanotechnology Lett.* **4**(10) 1015-1022(8)

[17] Frayssinet E, Knap W, Lorenzini P, Grandjean N et al 2000 High electron mobility in AlGaN/GaN heterostructures grown on bulk GaN substrates *Appl. Phys. Lett* **77** 2551-2553

[18] Kastalsky A A and Shur M S 1981 Conductance of Small Semiconductor Devices *Solid State Commun.* **39**(6) 715-718

[19] Dmitriev A P and Shur M S 2006 Ballistic admittance: Periodic variation with frequency *Appl. Phys. Lett.* **89** 142102

[20] Rumyantsev S, Shur M S, Knap W et al 2004 1/f noise and "ballistic mobility" in GaN/AlGaN heterostructure field effect transistors in high magnetic fields, in Noise in Devices and Circuits II *Proc. SPIE* **5470** 277-285

[21] Dmitriev A P, Kachorovskii V and Shur M S 2001 Plasma wave instability in gated collisionless two-dimensional electron gas *Appl. Phys. Lett* **79**(7) 922-924

[22] Kachorovskii V Yu, Roumyantsev S L, Knap W and Shur M 2013 Performance Limits for Field Effect Transistors as Terahertz Detectors *http://arxiv.org/abs/1302.7089*

[23] El Fatimy A, Dyakonova N, Teppe F et al 2006 Terahertz detection by GaN/AlGaN transistors *Electron. Lett.* **42**(23) 1342-1343

[24] Muravjov A V, Veksler D B, Popov V V, Polischuk O, Hu X, Gaska R, Pala N, Saxena H, Peale R E, Shur M S 2010 Temperature dependence of plasmonic terahertz absorption in grating-gate GaN HEMT structures *Appl. Phys. Lett.* **96** 042105

[25] Popov V V, Shur M S, Tsymbalov G M and Fateev D V 2007 Higher-order plasmon resonances in GaN-based field-effect transistor arrays *Int. J. High Speed Electron. and Systems* **17**(3) 557-566

Journal of Physics: Conference Series **486** (2014) 012026 doi:10.1088/1742-6596/486/1/012026

Superconducting integrated terahertz receivers

V P Koshelets[1], A B Ermakov[1], L V Filippenko[1], M Yu Fomisky[1], N V Kinev[1], O S Kiselev[1], A de Lange[2], G de Lange[2], K E Rudakov[1], M Yu Torgashin[1], V L Vaks[3], J Yuan[4] and H B Wang[4]

[1]Kotel'nikov Institute of Radio Engineering and Electronics of RAS, Moscow, Russia
[2]SRON Netherlands Institute for Space Research, 9700 AV Groningen, the Netherlands
[3]Institute for Physics of Microstructure RAS; Nizhny Novgorod; Russia
[4]National Institute for Materials Science, Tsukuba 3050047, Japan

E-mail: valery@hitech.cplire.ru

Abstract. A Superconducting Integrated Receiver (SIR) comprises on one chip all elements needed for heterodyne detection. Light weight and low power consumption combined with nearly quantum limited sensitivity and a wide tuning range of the superconducting local oscillator make SIR a perfect candidate for many practical applications.

1. Introduction and background

A Superconducting Integrated Receiver (SIR) [1] comprises on one chip a low-noise SIS mixer with quasioptical antenna, an Flux-Flow Oscillator (FFO) acting as a Local Oscillator (LO) and a second SIS harmonic mixer (HM) for the FFO phase locking (see figure 1). The concept of the SIR looks very attractive for many practical applications due to its compactness and the wide tuning range of the FFO. Presently, the frequency range of most practical heterodyne receivers is limited by the tunability of the local oscillator, typically 10-15% for a solid-state multiplier chain. In the SIR the bandwidth is determined by the SIS mixer tuning structure and the matching circuitry between the SIS and the FFO. A bandwidth up to 35% has been achieved with a twin-junction SIS mixer design. All components of the SIR microcircuits are fabricated in a high quality Nb based tri-layer on a Si substrate. The receiver chip is placed on the flat back surface of the silicon lens, forming an integrated lens-antenna.

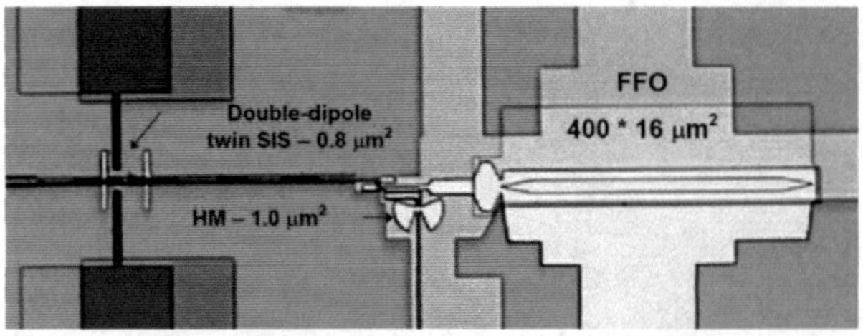

Figure 1. Central part of the SIR chip with antenna, twin SIS-mixer and harmonic mixer for FFO phase locking.

Content from this work may be used under the terms of the Creative Commons Attribution 3.0 licence. Any further distribution of this work must maintain attribution to the author(s) and the title of the work, journal citation and DOI.
Published under licence by IOP Publishing Ltd

2. Results

Continuous tuning of the phase-locked local oscillator has been realized at any frequency in the range 300-750 GHz. The output power of the FFO is sufficient to pump the matched SIS mixer in a wide frequency range and can be electronically adjusted. The FFO free-running linewidth has been measured between 0.3 and 5 MHz; resulting in the spectral ratio of the phase-locked FFO above 70% over the range. As a result of receiver's optimization the DSB noise temperature was measured below 100 K that is about 4 hf/kB; the spectral resolution is well below 1 MHz.

All these achievements enabled the development of a 450 - 650 GHz integrated receiver for the atmospheric-research instrument TELIS (TErahertz and submillimeter LImb Sounder) [2] - the balloon-borne instrument for the detection of spectral emission lines of stratospheric trace gases that have their rotational transitions at THz frequencies. Diurnal cycle of ClO has been observed; the BrO line with a level of only 0.3 K was isolated and clearly detected.

Capability of the SIR for high resolution spectroscopy has been successfully proven also in a laboratory environment by gas cell measurements. The possibility to use SIR devices for the medical analysis of exhaled air has been demonstrated. Many medically relevant gases have spectral lines in the sub-terahertz range and can be detected by a SIR-based spectrometer.

Recently the SIR was successfully implemented for the first spectral measurements of THz radiation emitted from intrinsic Josephson junction stacks (BSCCO mesa) in the frequency range 585 – 735 GHz; linewidth as low as 7 MHz has been recorded in the high bias regime. The phase-locked SIR has been used not only for detection of the BSCCO oscillator emission, but also for the locking of the oscillator under the test. The IF signal down-converted by the SIR is actually a convolution of the BSCCO oscillator signal and stable phase-locked SIR LO. This signal is applied to the PLL, where phase of the signal is compared with phase of the stable reference; the error signal is returned back to the BSCCO oscillator to control it phase. That is the first, but very important step towards development of fully HTc phase-locked local oscillator.

Nowadays the SIR is probably the most functionally complex fully superconducting device that was already successfully implemented for practical applications. In particular the SIR is very attractive for future airborne and space-borne missions as well as for analysis of the breathed out air at medical survey and for security monitoring.

Acknowledgments

The work was supported in part by the RFBR and the Ministry of Education and Science of the Russian Federation.

References

[1] Koshelets V P and Shitov S V 2000 Integrated Superconducting Receivers *Supercond. Sci. Technol.* **13** R53-R69
[2] Gert de Lange et al 2010 Development and Characterization of the SIR Channel of the TELIS Atmospheric Sounder *Supercond. Sci. Technol.* **23** 045016

Journal of Physics: Conference Series **486** (2014) 012027 doi:10.1088/1742-6596/486/1/012027

BWO based THz imaging system

Artem N Perov, Kirill I Zaytsev, Irina N Fokina, Valeriy E Karasik, Egor V Yakovlev and Stanislav O. Yurchenko

Bauman Moscow State Technical University, 2nd Baumanskaya str., 5, Moscow, 105005, Russia

E-mail: kirzay@gmail.com

Abstract. Terahertz (THz) imaging is a relatively new field of optical engineering and physics. Despite it has been already used in such directions as biomedical imaging, non-destructive evaluation of materials, and security detection of concealed objects, there are many problems in the development of them, connected with low efficiency of imaging sources and detectors. A new type of THz imaging system was implemented, based on backward wave oscillator THz source. Its efficiency of detecting concealed objects was shown.

1. Introduction

As THz radiation has high penetration depth in dielectric materials [1],[2],[3]], it can be used for detection of different concealed objects, especially within the security problems. This idea was developed during the last several years [4]-[11]. Depending on the source of THz radiation there are two promising types of imaging: using pulse source within time of light imaging scheme (see for example [12],[13]) as it is usually utilized in THz time-domain spectroscopy, and continuous THz systems [12],[14],[15]].

In pulse imaging systems, there is a big disadvantage of low power efficiency [16]-[18]. Other negative factor is long time of scanning (depended on the desirable resolution). In continuous imaging systems, as a source of THz radiation Gunn diode, photomixing scheme or backward wave oscillator (BWO) are often used [12],[19]. Unfortunately, detectors used in these systems have low sensitivity. Thus it is essential to use more powerful THz sources.

In this work, we suggest a new THz imaging system scheme, based on BWO 0.25THz source and heterodyne Schottky diode detector module. For increasing a source power input, we have studied BWO output beam profile and deviation and have implemented beam correction module.

The suggested imaging system was tested during a set of experiments. Estimation of the obtained images of concealed metal objects shows the ability of this system to detect such objects under different types of fabric and clothes.

2. THz imaging system description

The suggested scheme of active THz imaging system (figure 1) uses BWO continuous source of 0.25 THz radiation with average output power 50 mW. BWO generates a THz beam with rather complicated profile (figure 2). Arbitrary peaks, inhomogeneity, and ring-structure are well noticed. To increase illumination power, it is necessary to reduce inhomogeneity and make divergence of output beam more smooth. For this purpose a special beam integrator was developed (figure 3).

Content from this work may be used under the terms of the Creative Commons Attribution 3.0 licence. Any further distribution of this work must maintain attribution to the author(s) and the title of the work, journal citation and DOI.
Published under licence by IOP Publishing Ltd

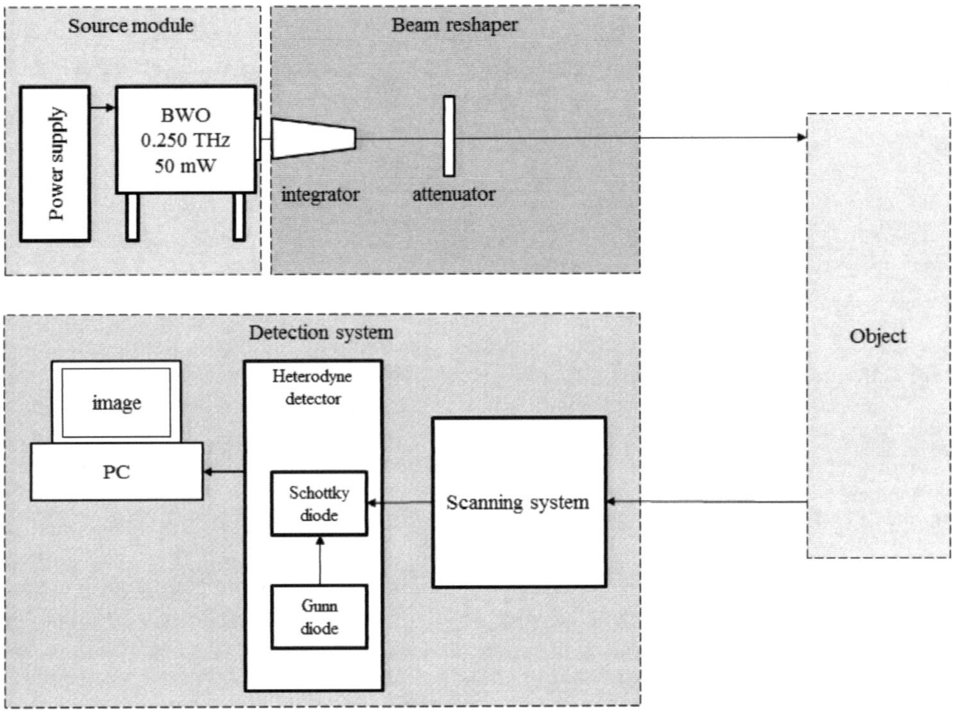

Figure 1. BWO based THz imaging system scheme.

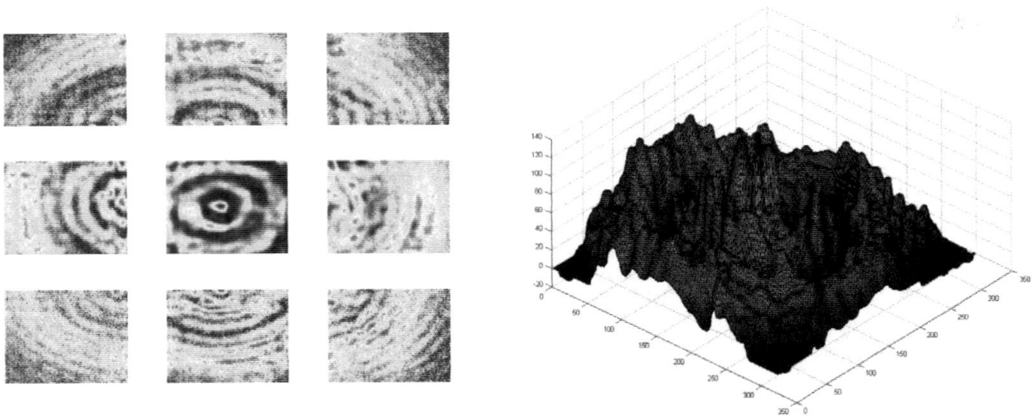

a b

Figure 2. BWO beam registered by pyroelectical matrix beam profiler: a) is the result of scanning in lateral direction, b) 3D beam profile.

Journal of Physics: Conference Series **486** (2014) 012027 doi:10.1088/1742-6596/486/1/012027

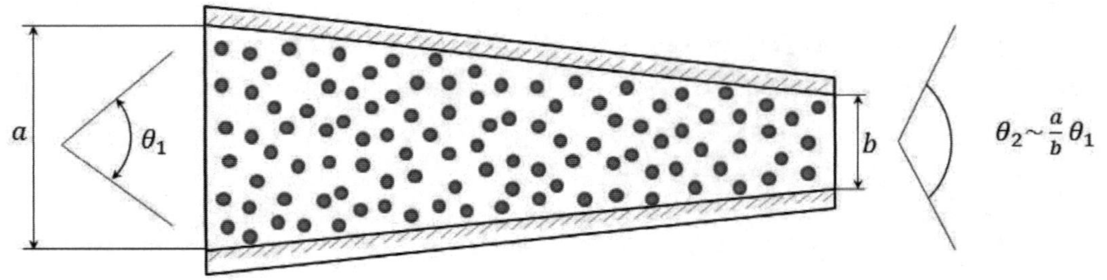

Figure 3. Beam integrator: internal surfaces are reflecting plane mirrors; beam divergence increases in k times; partially internal space is filled by the scattering nonabsorbent medium.

The internal surface of this integrator is reflecting. The ratio between input and output apertures is $\alpha = a/b$. Integrator is placed close to the BWO output hole, and thus the beam divergence increases in $k = \alpha$ times. Partially the internal integrator space is filly by dielectric scattering medium for making output beam homogeneous. Right after the integrator, an attenuator is placed for reducing output illuminating power. The THz detector is based on the Schottky diode heterodyne scheme with optical mechanical scanning system. In heterodyne detector, the radiation reflected from the object is mixed with radiation from the reference heterodyne source – Gunn diode. The resulted signal is visualized on the PC monitor. Image of the object is normalized to the brightest object point. If the BWO illuminating beam is reflected from highly reflecting object, then it is possible to obtain only one bright point on the image. To prevent this situation, attenuator is used.

3. Experimental results

During the experiment, images of the test-objects were obtained by the suggested THz imaging system. We used aluminum foil stripes on paper: paper is highly transparent at 0.25 THz, while foil is highly reflecting. This set consisted of 6 objects with different stripes frequency 0.031, 0.050, 0.0625, 0.083, 0.125, 0.250 cm^{-1} (figure 4).

Figure 4. Set of the test-objects.

The described test-objects were observed on man under different clothes: coat, jacket, and synthetic fabric. The distance between the object and the imaging system was 3.5 meters. From the figure 5 it can be clearly seen, that all these objects were detected by our visual system, but the resolution is limited for the last test-object due to the scattering properties of clothes.

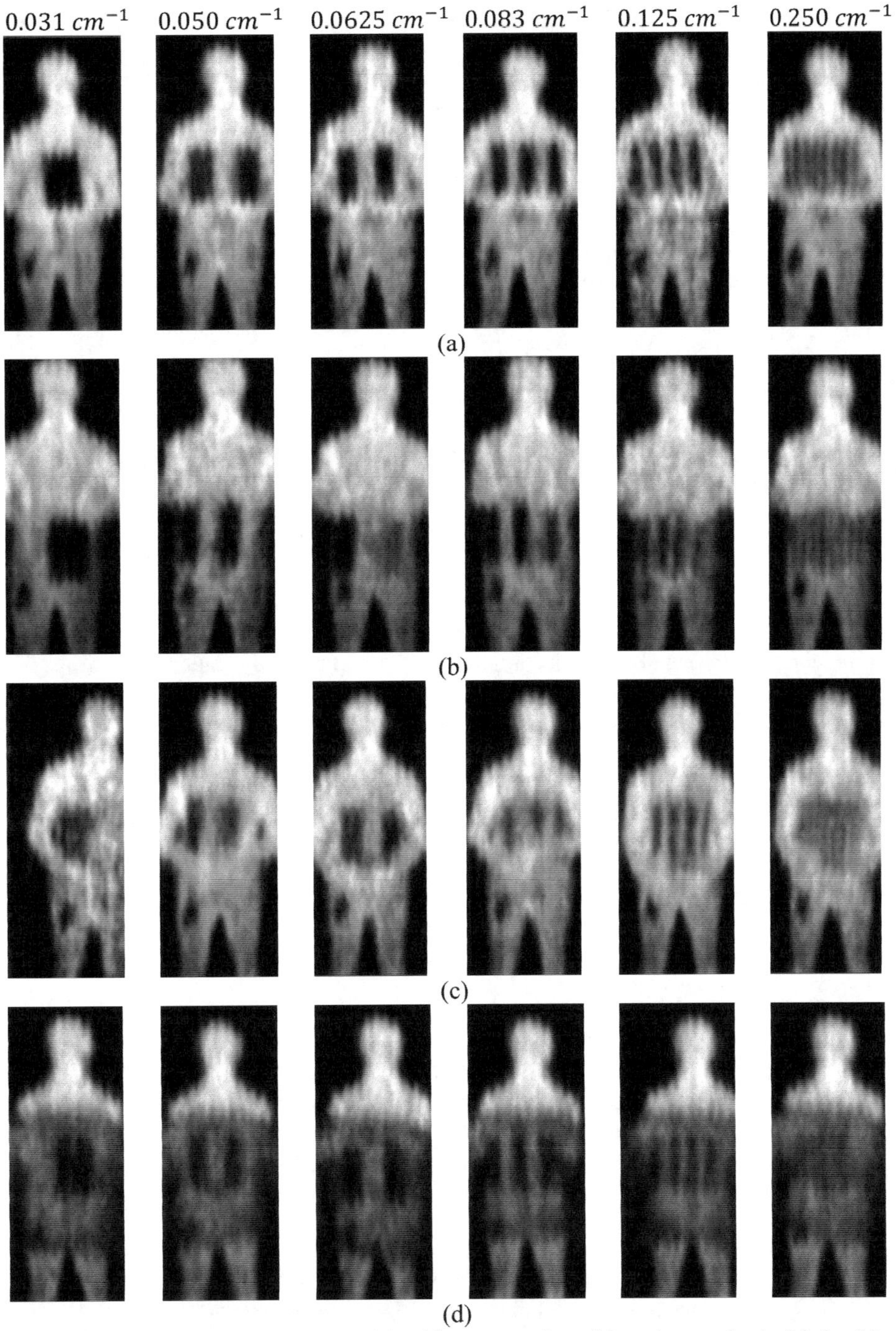

Figure 5. Images of test-objects: (a) without covering, (b) under synthetic fabric, (c) under jacket, (d) under coat.

4. Conclusion

THz imaging system based on BWO source and heterodyne detector was developed. The BWO beam quality was increased without significant power losses. This system can be effectively used for finding concealed objects: metal things, weapon, knifes. Results of the experiment show the ability to resolve little parts of objects under thin fabric up to 0.250 sm^{-1} and under thick clothes (coats) up to 0.125 sm^{-1}.

References

[1] Liu J, Fan W, Xue B 2011 Study on the transmission characteristics of terahertz pulse through packing materials *Chin. Opt. Lett.* **9** S10202

[2] Peng B, Ding T and Wang P 2012 Propagation of polarized light through textile material *Appl. Opt.* **51**(26) 6325-6334

[3] Dieltlein C r, Bjarnason L E, Grossman E N and Popovic Z 2008 *Proc. SPIE* **6948** 69480E

[4] Hoshina H, Sasaki Y, Hayashi A, Otani C and Kawase K 2009 Noninvasive mail inspectation system with terahertz radiation *Appl. Spectroscop.* **63**(1)

[5] Sinyukov A, Zorych I, Michalopoulou, Z-H, Gary D, Barat R, Federici J F 2008 Detection of explosives by terahertz synthetic aperture imaging – focusing and special classification *Comptes Rendus Physique* **9**(2) 248-261

[6] Miles R E 2007 *Terahertz frequency detection and identification of materials and objects* (New York: Springer) 251-323

[7] Leahy-Hoppa M R, Fitch M J, Zheng X, Hayden L M, Osiander R 2007 Wideband terahertz spectroscopy of explosives *Chem. Phys. Lett.* **434** 227-230

[8] Federici J F, Schulkin B, Huang F, Gary D, Barat R, Oliveira F and Zimdars D 2005 THz imaging and sensing for security applications – explosives, weapons and drugs *Semicond. Sci. Technol.* **20** S266–S280

[9] Grossman E et al 2010 Passive terahertz camera for standoff security screening *Appl. Opt.* **49**(19)

[10] Cooper K B et al 2010 Fast, high-resolution terahertz radar imaging at 25 meters *Proc. SPIE* **7671** 76710Y

[11] Kemp M C 2006 *Proc SPIE* **6402** 94020D

[12] Karpowicz N, Zhong H, Xu J, Lin K.-I, Hwang J-S, Zhang X-C 2005 Comparison between pulsed terahertz time-domain imaging and continuous wave terahertz imaging *Semicond. Sci. Technol.* **20** S293–S299

[13] Kawase K, Ogawa Y, Watanabe Y 2003 Non-destructive terahertz imaging of illicit drugs using spectral fingerprints *Opt. Express* **11**(20)

[14] Karpowicz N, Dawes D, Perry M J, Zhang X-C 2006 Fire Damage on Carbon Fiber Materials Characterized by THz Waves *Proc. SPIE* **6212** 62120G

[15] Zhang H, Mitobe K, Kabir M, Suzuki M, Mitobe Y, Habuchi T, Yoshimira N 2012 Improvement of Continuous-Wave Terahertz Imaging System for Cellulose Acetate Membrane Electrophoresis *IEEJ Trans. Elect. Electron. Eng.* **7**(S1): S126–S131

[16] Ryzhii V, Otsuji T, Ryzhii M, Leiman V G, Yurchenko S O, Mitin V, Shur M S 2012 Effect of plasma resonances on dynamic characteristics of double graphene-layer optical modulator *J. Appl. Phys.* **112**(10) 104507

[17] Ryzhii V, Otsuji T, Ryzhii M, Ryabova N, Yurchenko S O, Mitin V, Shur M S 2013 Graphene terahertz uncooled bolometers *J. Phys. D: Appl. Phys.* **46**(6) 065102

[18] Ryzhii V, Satou A, Otsuji T, Yurchenko S O, Ryzhii M, Ryabova N, Shur M S 2012 Graphene-based electro-optical modulator: Concept and analysis *IMFEDK 2012 - 2012 International Meeting for Future of Electron Devices, Kansai* **6218573** 44-45

[19] Chen G, Pei J, Yang F, Zhou X Y, Sun Z L and Cui T J 2012 Terahertz-Wave Imaging System Based on Backward Wave Oscillator *IEEE Trans. THz Sci. Technol.* **2** (5)

Journal of Physics: Conference Series **486** (2014) 012028 doi:10.1088/1742-6596/486/1/012028

Hot-electron micro&nanobolometers based on low-mobility 2DEG for high resolution THz spectroscopy

Vladimir Mitin[1], Victor Pogrebnyak[1], Michael Shur[2], Remis Gaska[3], Boris Karasik[4] and Andrei Sergeev[1]

[1]University at Buffalo, the State University of New York, USA
[2]Rensselaer Polytechnic Institute, ECSE Department, USA
[3]Sensor Electronic Technology Inc., Columbia, USA
[4]Jet Propulsion Laboratory, California Institute of Technology, USA

E-mail: asergeev@buffalo.edu

Abstract. The results on design, fabrication, and characterization of a hot-electron bolometer (HEB) based on the low-mobility two-dimensional electron gas (2DEG) in a AlGaN/GaN heterostructures show that our HEBs have high coupling to incident THz radiation due to the Drude absorption. Significant heating by THz radiation is realized due to the small value of the electron heat capacity. A low contact resistance achieved in our devices ensures that the THz radiation couples primarily to the 2DEG. Due to the small electron momentum relaxation time, the real part of the 2DEG sensor impedance is ~50-100 Ω, which provides a good impedance match between the sensor and antenna. Currently the room temperature responsivity of our devices reaches ~ 0.04 A/W at 2.55 THz along with a noise equivalent power ~ 5 nW/Hz$^{1/2}$at room temperature.

1. Introduction and background

THz heterodyne receivers currently use Schottky diodes, Superconductor-Insulator-Superconductor (SIS), plasma electronics detectors, and superconducting hot-electron-bolometers.

For a long time, Schottky diode mixers were used in receivers for frequencies up to THz range. The main advantage of these mixers is the possibility of room temperature operation. However, the key problem is that the Schottky diode requires high LO power for its operation as a mixer, in addition, they have rather poor sensitivities above the 1-2 THz range.

Superconductor-Insulator-Superconductor (SIS) tunnel junctions operate at liquid helium temperatures. However, these devices have high sensitivity only at frequencies below the superconducting energy gap. Above the superconducting gap frequency (e.g. ~1.2 THz for NbTi), the performance of SIS mixers reduces dramatically.

Developed a decade ago, superconducting hot electron bolometer (HEB) mixers are currently the devices of choice for the frequencies above 1 THz. The heterodyne mixing process is based on temperature dependence of the bridge resistively in the transition (resistive) state. The IF bandwidth is determined by the inverse electron cooling time, which can be varied from the electron-phonon relaxation time in long wires to the hot-electron outdiffusion time in short bridges. Currently, NbN devices show the best sensitivity, such as a noise temperature, T_n, of 750 K at 1.9 THz and 950 K at 2.5 THz, in combination with an IF gain bandwidth of ~ 6 GHz. However, they require deep cooling, down to helium temperatures. Cryogenics requirements are a main obstacle for these devices.

Content from this work may be used under the terms of the Creative Commons Attribution 3.0 licence. Any further distribution of this work must maintain attribution to the author(s) and the title of the work, journal citation and DOI.
Published under licence by IOP Publishing Ltd

Two-dimensional electron gas in semiconductor heterostructures was identified as a promising medium for heterodyne mixing in early 90s [1]. Up until now the only material, which has been pursued in view of its availability, is AlGaAs/GaAs. However, it is currently understood, that parameters of AlGaAs/GaAs heterostructures (high mobility and low carrier concentration) do not allow for achieving of a reasonable coupling to THz radiation above ~1THz [2].

2. Results

We investigate capabilities of a novel fast (picosecond) THz detector for use in advanced heterodyne receivers, operating from liquid nitrogen to room temperatures. The detector operation is based on the effect of electron heating of low mobility two-dimensional electron gas (2DEG) in micron/submicron patterned GaN heterostructures [3].

The GaN heterostructures were grown by memory enhanced metal-organic chemical vapor deposition (MEMOCVD®) on two inch diameter (0001) sapphire substrates (see Fig. 1). First, a 200 nm undoped AlN layer was grown on the substrate, this layer was grown to reduce surface roughness on the sapphire substrate and to prevent propagating of defects upward. After that, a 2.5 μm non-intentionally doped GaN buffer layer was grown. Then a 1 nm high Al-content AlGaN spacer layer and a 9.5 nm Si-doped $Al_{0.88}Ga_{0.12}N$ top layer were deposited. Finally, the samples were covered with SiO_2 to avoid oxidation of Al in the top layer. The room temperature electron mobility and concentration were extracted using Hall and Van-der-Paw measurements.

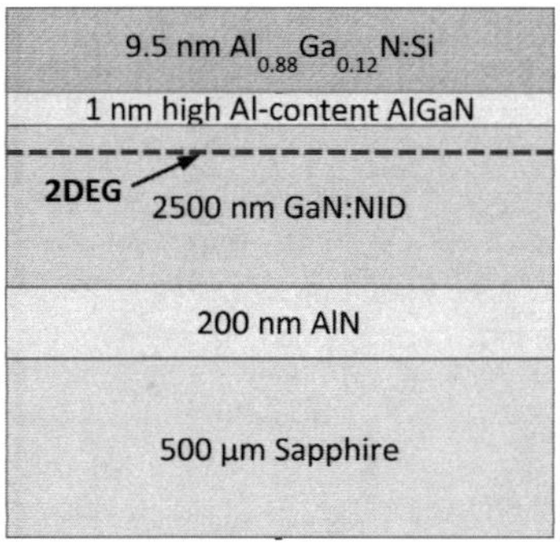

Figure 1. Schematic of AlGaN/GaN heterostructures with 2DEG.

A combination of the short electron mean free path with high electron density of 2DEG provides a sheet resistance of 200 - 300 Ω/□ that meets the requirements for the impedance matching between conventional THz antennas and a 2DEG sensor. Three different wafers with a different level of doping were grown for our devices.

The wafers were processed into mesa structures by photolithography and dry etching in a $SiCl_4$/Argon mixture plasma. Prior to the dry etching process, the samples were prepared by organic cleaning procedure, using trichloroethylene, acetone and methanol. A Ti/Al/Ti/Au metal stack having a thickness of 25/70/50/100 nm was deposited by using an e-beam evaporator after the etching process, followed by rapid thermal anneal (RTA) at 850 °C in a N_2 ambient to form ohmic contacts. Finally, an antenna made of a 200 nm Ni/Au was deposited on the top of the structure. The antennas deposited on the ohmic contacts were used also as extended contact pads by the external read-out circuit.

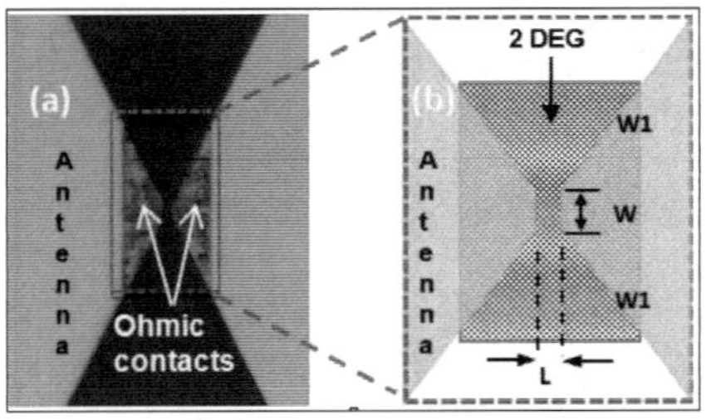

Figure 2. Image and schematic diagram of Devices 1 and 2 with ohmic coupling. The apexes of the antenna and ohmic contacts have an identical shape. The ohmic contacts are shown in dark color.

Our devices discussed here have three types of broadband antennas covering a range of frequencies from 1 THz to 4 THz: (1 and 2) a regular bow-tie antenna (figure 2), (3) a spiral antenna with ohmic coupling (figure 3), (4) a bow-tie antenna with a capacitive coupling (figure 4). In the devices with ohmic coupling, an antenna was deposited on the top of the ohmic contacts, which transfer current from the antenna to the 2DEG. Device 1 had the mesa area of 40×40 μm^2 (marked by red dash line in figure 2a) and the channel area $L \times W = 2 \times 2$ μm^2 (see figure 2b). Device 3 with ohmic contacts had a broadband spiral antenna with 2DEG channel of 2×10 μm^2 (figure 3).

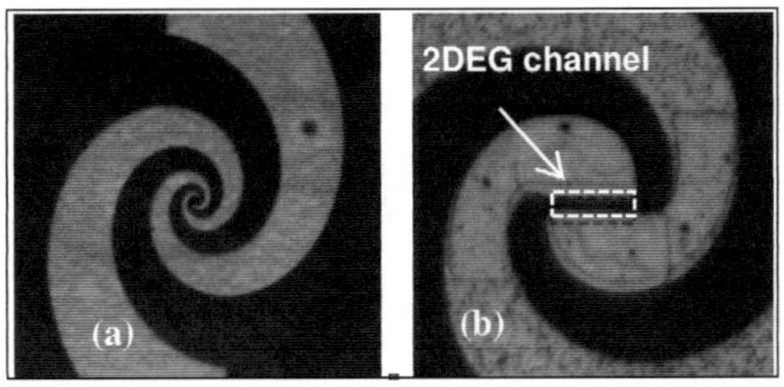

Figure 3. (a) Image of Device 3 with ohmic coupling, (b) the apexes of the spiral antenna forms a 2×10 μm^2 2DEG channel.

Device 4 used capacitive coupling and spit-gate patterning of the sensor (figure 4) with the bow-tie antenna not having ohmic contacts to the 2DEG and with the mesa tapered from the contact sides towards the antenna terminals (figure 4b). The narrow 2 μm × 4 μm channel between the bow-tie antenna apexes created a current constriction area. The THz electric field induced in the gap between the antenna apexes penetrated into 2DEG layer and heated the electron gas in the channel. The design with capacitive coupling allows one to avoid significant technological problems related to fabrication of low-ohmic contacts. In addition, the antenna's bows may be used as side split-gates to decrease the channel width by applying a negative potential to the gate. This design allowed us to control and optimize the channel width.

The results of the measurements for Devices 1 and 2 are shown in figure 5. As seen, the Device 1 responsivity reaches 14 mA/W at a bias of 10 mA. It is approximately 1.5 times greater than that for Device 2, which is in good agreement with the interpretation in terms of the parameter $\lambda^2/(L \times W)$, since the ratio of the areas of these devices equal to 2. The small difference can be accounted for the different antenna impedances. For the polarization along the arrow on Fig. 4b, the photoresponse is

Journal of Physics: Conference Series **486** (2014) 012028 doi:10.1088/1742-6596/486/1/012028

five times lower because the antenna did not develop voltage for this polarization of THz radiation, and the entire response, was due to the direct absorption of THz radiation by the 2DEG.

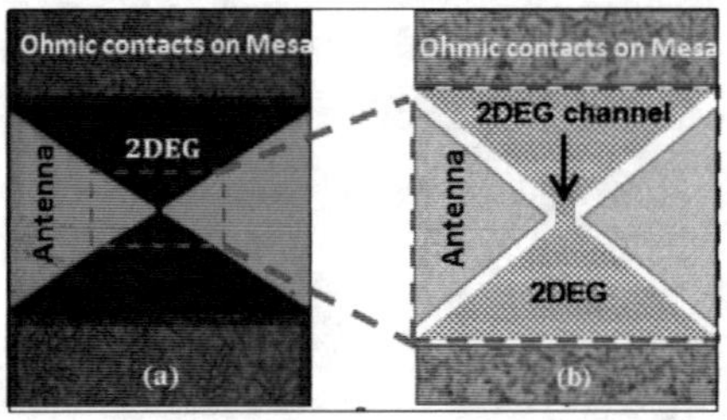

Figure 4. Image (a) and schematic diagram (b) of Device 4 with capacitive coupling, (b) the mesa is shaped in a form having a narrow 2DEG channel between the apexes of the bow-tie antenna.

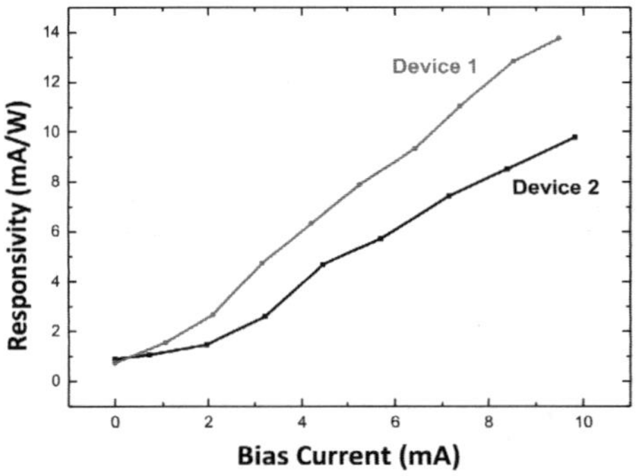

Figure 5. Responsivity versus bias current for Devices 1 and 2.

The THz photoresponse of Device 3 was measured at a laser power of 58 mW/cm^2 using the chopping frequency of f = 1 kHz. The THz response of this device was three times greater than the response of Device 1 and reaches 40 mA/W at a bias current of 8 mA. The enhanced response may be explained by the small areas of the mesa and of the channel in this device.

The device 4 antenna had the flare angle of 60°, impedance of 85 Ω, and gain of 5 dB. The DC resistance of the device was 1.9 kΩ. The following experimental settings were chosen for the device characterization. The laser power was 50 mW/cm^2 and the chopping frequency was 355 Hz. For the polarization along the arrow on figure 4b, the photoresponse was two times smaller than for the polarization in the direction of the antenna that is perpendicular the arrow on figure. 4b. For the first polarization, the antenna does not enhance coupling to 2DEG. However, the 2DEG itself may operate as an antenna, because the 2DEG channel had a shape of the bow-tie antenna with a flare angle of 120°. The results of the measurements for the second polarization of THz radiation are shown in figure 6. As it is seen from figure 6, the response of Device 4 with capacitive coupling is similar to those shown in figure 5.

Journal of Physics: Conference Series **486** (2014) 012028 doi:10.1088/1742-6596/486/1/012028

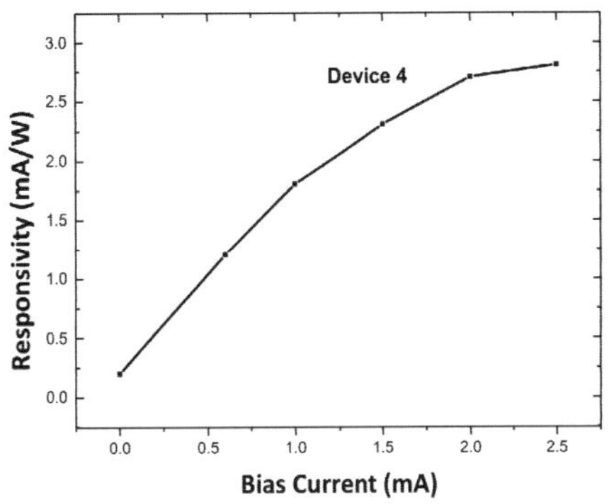

Figure 6. Responsivity versus bias current for Device 4 with capacitive coupling.

We conducted first measurements of the 2DEG HEB mixer frequency bandwidth. Because the bandwidth of a HEB mixer does not depend on frequencies of the mixing waves, we used available equipment: two Gunn diodes, where one was used as a local oscillator and the other diode as a source. The local oscillator (LO) frequency was tunable between 81 and 98 GHz and the signal source (RF) was fixed at 81 GHz. The LO and RF powers were brought together through a 6 dB directional coupler followed by a calibrated attenuator. Since all measurements were done at 300 K the cryostat was used only for sample mounting purpose, but in the next experiments with QCL we may use it for low temperature experiments. The LO and RF signals were coupled to the 2DEG mixer by placing the mixer at the open end of the waveguide. Figure 7 shows the normalized IF power versus an IF frequency measured at 300K for Device 3. The 3 dB bandwidth defined by $BW = 1/2\pi\tau$ was found to be equal to 2 GHz that correspond to the maximum bandwidth achievable with our setup. This result shows that the electron cooling time is shorter than 100 ps.

In summary, the feasibility of 2DEG GaN hot electron bolometer (HEB) for THz heterodyne sensing in the range 1-4 THz has been demonstrated. Hot-electron bolometers based on the low-mobility two-dimensional electron gas (2DEG) in AlGaN/GaN heterostructures have been designed and fabricated. Electrical, optical, thermal, and THz characterizations of our devices show that the low mobility ($\mu \approx 2000$ cm^2/Vs) and the corresponding short electron momentum relaxation time provide a strong coupling between the 2DEG and incident THz radiation due to the Drude absorption. A high electron concentration ($n \sim 10^{13}$ cm^{-2}) in 2DEG structure provides the real part of the 2DEG sensor impedance of ~50-300 Ω, which allows for impedance matching between the sensor and conventional broadband THz antennas. Our estimates show that the responsivity and NEP of our devices can be significantly improved by 1) further reduction the sensor area, 2) growing GaN structures with increased doping to further increase carrier concentration and reduce momentum relaxation rate. This will improve the impedance match between the antenna and 2DEG sensor at THz frequencies.

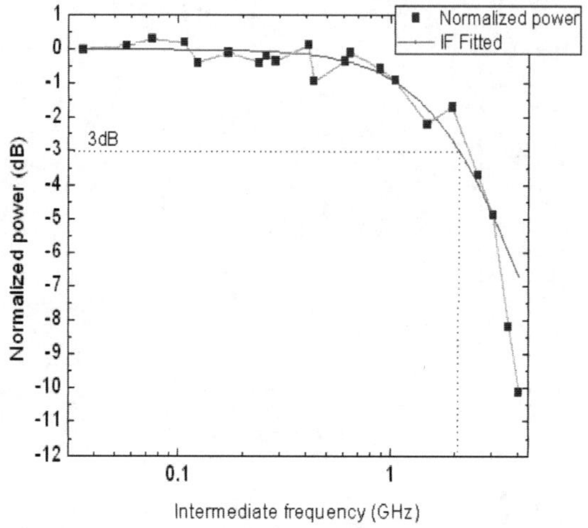

Figure 7. Normalized IF output power versus an intermediate frequency for the mixer prototype.

References

[1] Yang J-X, Agahi F, Dai D, Musante C F, Grammer W, Lau K M and Yngvesson K S 1993 Wide-bandwidth electron bolometric mixers: A 2DEG prototype and potential for low noise THz receivers *IEEE Trans. Microwave Theory Tech. MTT* **41** 581

[2] Lee M, Pfeiffer L N and West K W 2002 Ballistic cooling in a wideband two-dimensional electron gas bolometric mixer *Appl. Phys. Lett.* **81** 1243–1245

[3] Choi J K, Mitin V, Ramaswamy R, Pogrebnyak V A, Pakmehr M, Muravjov A, Shur M, Gill J, Mehdi I, Karasik B S and Sergeev A 2013 THz Hot-Electron Micro-Bolometers Based on Low-Mobility 2DEG in GaN Heterostructures *IEEE Sensors Journal* **13** 80

IOP Publishing
doi:10.1088/1742-6596/486/1/012029

Frequency dispersion and damping mechanisms of terahertz plasmons in graphene transistor structures

Akira Satou[1], V Ryzhii[1], Vladimir V Mitin[2], Fedir T Vasko[2] and Taiichi Otsuji[1]

[1]Research Institute of Electrical Communication, Tohoku University, Japan
[2]Department of Electrical Engineering, University at Buffalo, USA

E-mail: a-satou@riec.tohoku.ac.jp

Abstract. We develop a numerical model for simulation of plasmons in graphene with grating-gate structures, based on the Boltzmann equation and self-consistent Poisson equation. Using the model, we study the effect of coupling between plasmons in gated and ungated regions on the frequency tunability by the gate voltage and the damping of plasmons due to carrier scattering and due to source and drain contacts.

1. Introduction and background

Plasmons in graphene have attracted much attention recently for applications to compact, frequency-tunable terahert (THz) sources and detectors that operate at room temperature [1]. One of the most important advantages of plasmons in graphene is the long momentum relaxation time in clean monolayer graphene and, hence, the weak damping of plasmons in the THz range. In this report, we study frequency dispersion and damping mechanisms of THz plasmons in graphene transistor structures by numerical simulation.

2. Results

We consider a graphene channel with periodic single-grating gates shown in figure 1 with electron doping concentration 10^{12} cm^{-2}. Our numerical model is based on the Boltzmann equation for electron and hole transport with the Poisson equation for self-consistent electric field. We adapt the so-called weighted essentially-nonoscillatory scheme to solve the Boltzmann equation, whereas the finite-element method is used for the Poisson equation. Figure 2 shows the gate-voltage dependence of frequency of the fundamental plasmon mode. It clearly shows that the frequency is in the THz range and is tunable by the gate voltage. The frequency is higher in the structure with L_g = 1000 nm than with L_g = 2000 nm, whereas the voltage tunability of the frequency is wider in the latter than in the former. These are related to the coupling of plsmons in the gated and ungated regions.

Figure 3 shows the dependence of the damping rates due to the acoustic-phonon scattering at room temperature on the electron concentration, as well as damping rates due to short- and finite-range disorder scattering. The damping rate in disorder-free graphene can be as low as 1.3×10^{11} s^{-1}. The damping rates due to disorder scattering can be on the the same order if the the point-defect concentration n_{DS} is as low as 10^{12} cm^{-2} and the correlation length l_{DF} for the finite-range disorder, characterizing the inhomogeneity of graphene, is longer than 100 nm. In addition to the damping caused by carrier scattering, we studied the damping caused by the presence of equilibrium source and drain contacts, which was reported for compound semiconductor two-dimensional electron systems [2]. It is demonstrated that the damping rate is proportional to v_F/L, where $v_F = 10^8$ cm/s. This damping

Content from this work may be used under the terms of the Creative Commons Attribution 3.0 licence. Any further distribution of this work must maintain attribution to the author(s) and the title of the work, journal citation and DOI.
Published under licence by IOP Publishing Ltd

Journal of Physics: Conference Series **486** (2014) 012029 doi:10.1088/1742-6596/486/1/012029

is caused by the leakage of the oscillation energy of plasmons to the equilibrium electron reservoirs, i.e., source and drain contacts.

In conclusion, we have shown that the frequency of plasmons in graphene transistor structures can be widely tuned in the THz range and the damping rate can be as low as 10^{11} s^{-1} at room temperature in disorder-free graphene, suggesting the ultimate performance of graphene-based THz plasmonic devices.

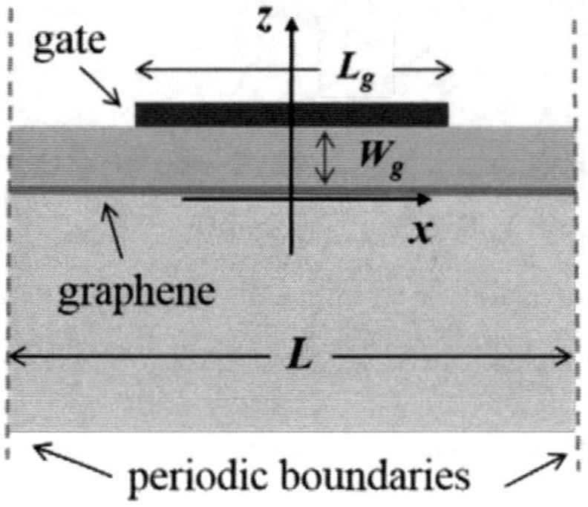

Figure 1. Schematic view of the single-grating-gate structure.

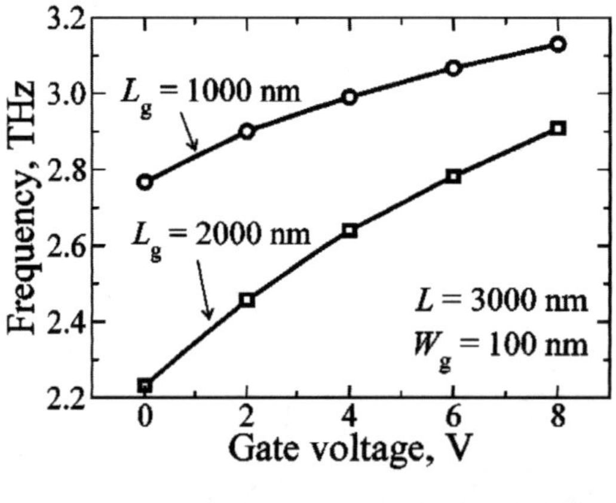

Figure 2. Gate-voltage dependence of the fundamental plasmon frequency in the structure with different gate lengths.

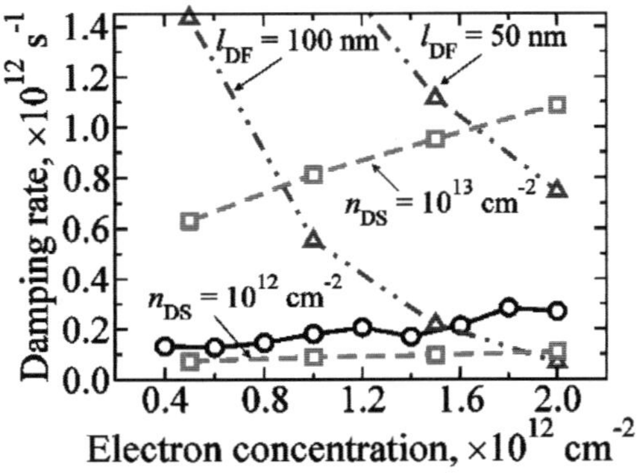

Figure 3. Damping rates caused by the acoustic-phonon, short-range disorder, and finite-range disorder scatterings (solid, dashed, and dashed-dotted lines, respectively).

References

[1] Ryzhii V 2006 *JJAP* **45** L923

[2] Satou A, Ryzhii V, Mitin V and Vagidov N 2009 *Phys. Status Solidi B* **246** 2146

Journal of Physics: Conference Series **486** (2014) 012030 doi:10.1088/1742-6596/486/1/012030

Nano-lasers and nano-LEDs

Magnus Willander

Department of Science and Technology, Campus Norrköping, Linköping University SE-601 74 Norrköping, Sweden

E-mail: magnus.willander@liu.se

Abstract. One-dimensional nanostructures have been of great interest for photonic devices. In this talk I will give a review of the one-dimensional ZnO nanostructures for ultraviolet and visible regions. I will discuss earlier and last results regarding room temperature and low temperature spontaneous and stimulated emission. For lasers we will discuss both optically and electrically pumped lasers. For the LEDs particularly I will discuss the white light emission from LEDs. Substrates like graphene, paper, textile and semiconductors will be discussed. Finally, also results from other devices based on hybrid materials will be analyzed.

1. Introduction and background

Zinc oxide (ZnO) with its self-organized growth property combined with a direct wide band gap (3.34 eV), relatively large excitons energy (60 meV), and deep band emission covering the whole visible range constitute an interesting material for the development of different photonic devices. In addition, the family ZnO nanostructures is one of the richest known families and it can be grown at low temperature and hence the fabrication of these photonic devices can be realized on non-conventional substrates, like e.g. glass, plastic, paper, etc.

Figure 1. Scanning electron microscope (SEM) image of ZnO NRs grown on textile substrate using the low temperature chemical growth.

Content from this work may be used under the terms of the Creative Commons Attribution 3.0 licence. Any further distribution of this work must maintain attribution to the author(s) and the title of the work, journal citation and DOI.
Published under licence by IOP Publishing Ltd

In addition, ZnO is bio-safe, biocompatible, and possesses piezoelectric property with relatively high electro-mechanical coupling coefficient [1]. The low temperature growth of ZnO nanostructures is rather simple and can be achieved on the above mentioned non-conventional substrates. Figure 1 shows a scanning electron microscope (SEM) displaying a typical example of ZnO nanorods (NRs) of high crystalline quality grown on textile substrates.

2. Results

The growth of the ZnO NRs is achieved as follows: as a first step a conducting layer, typically a metal like silver is deposited to provide the bottom contact to the ZnO NRs. To improve the ZnO nanorods growth quality, distribution and density a seed layer is spun-coated and baked for 20 min at 250 °C. Then zinc nitrate hexa hydrate (Zn $(NO_3)_2.6H_2O$) was mixed with Hexamethyl tetra-mine (HMT) ($C_6H_{12}N_4$) in equal molar concentartion and substrates were placed in the solution for four hours at 90 °C. Since the grown ZnO NRs are n-type by intnetional doping, a p-type polymer is used, either before the seed deposition or on top of the ZnO NRs after the growth. The fabricated white light emitting diodes (LEDs) have emitted white light of high quality. Figure 2 show a ZnO NRs based light emitting didoe under measurement.

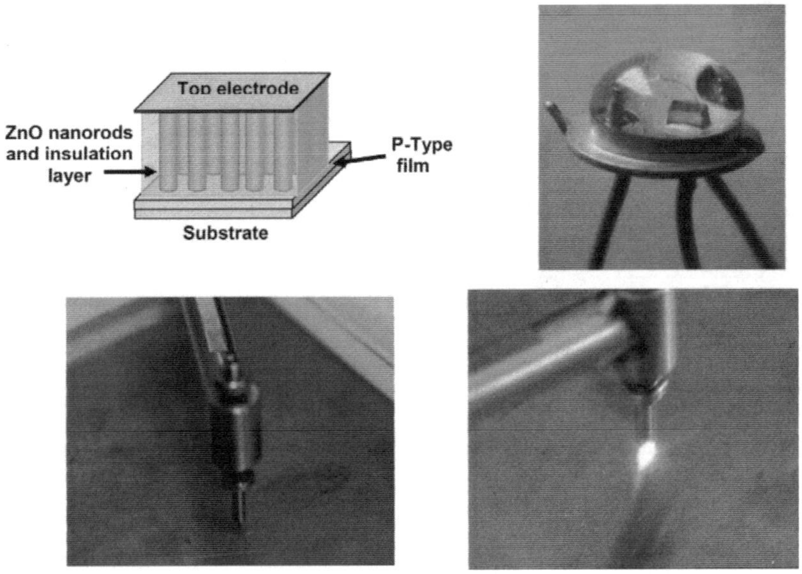

Figure 2. Schematic diagram showing one configuration of ZnO NRs based white light emitting diode (LEDs) (top left), a packaged white LED is shown also (top right side). The bottom two images show the white LED under measurement.

References

[1] Willander M et. al. 2009 Zinc oxide nanorods based photonic devices: recent progress in growth, light emititng diodes and lasers *Nanotechnology* **20** 332001

Journal of Physics: Conference Series **486** (2014) 012031 doi:10.1088/1742-6596/486/1/012031

The structure and spectral properties of two-dimensional dipole systems

Stanislav O Yurchenko and Nikita P Kryuchkov

Bauman Moscow State Technical University, 2nd Baumanskaya str., 5, Moscow, 105005, Russia

E-mail: st.yurchenko@mail.ru

Abstract. A molecular dynamics simulation of two-dimensional dipole systems was performed. The influence of particles dipole moment (PDM) on structures and spectral properties of the systems was analyzed. It was found that increase of PDM leads to qualitative changes in systems structure and spectrum. There is a critical ratio between temperature and PDM at which the system is polymerized. It was also shown that the spectral properties of the system are extremely sensitive to PDM.

1. Introduction

Processes occurring in condensed matter, especially in liquids, play an important role in modern engineering and technical issues [1]-[4]. However, despite the efforts made in this area complete theory about disordered states fluid and phase transition still does not exist. This is due to the fact that such theory must be able to describe the great variety of dynamic structures that can occur in condensed matter. Study of the liquid phase's properties, especially liquid water has received considerable attention. Water has several of anomalous properties. This is probably due to the fact that water molecules are connected by a network of hydrogen bonds can be formed in a very complex cluster groups, and various other structures. A large number of articles devoted to the study of structures that occur in liquid water (for example [5]-[7]).

Many studies in the field of condensed matter physics are performed by computer simulation in particular by molecular dynamics methods (MD). There are many models of water used in similar problems. For example review article [13] contains 46 different models of water. Most models of water are multisite. This allows more accurately reflect some of the properties of real water. However it also leads to a large amount of computer time necessary for calculation of interaction forces which imposes certain restrictions on the size of simulated system and simulation time.

There are a number of problems the study of which requires the simulation of large systems for long intervals of time. For example the existence of long-lived nanosized cavities in real water [11].

One of the simplest models of liquid is a system of particles interacting through Lennard-Jones potential (LJ) [8]. However, long-lived cavity in such systems is not observed. One of the features of water is the large dipole moment of the water molecule. Therefore there is a problem of investigating possibility that the existence of cavities is mainly due to the fact that the water molecules have a large dipole moment.

Therefore, it becomes necessary to perform a series of simulations of LJ systems the particles of

Content from this work may be used under the terms of the Creative Commons Attribution 3.0 licence. Any further distribution of this work must maintain attribution to the author(s) and the title of the work, journal citation and DOI.
Published under licence by IOP Publishing Ltd

which have a dipole moment. This model is probably closest to the SSD model [12] excluding torsion forces. In this study the influence of PDM values on structure and spectral properties of the system was analyzed.

2. Description of the modelling process

The system consists of uniform spherical particles, the centers of which interact through LJ potential:

$$U(r) = 4\varepsilon\left(\left(\frac{\sigma}{r}\right)^{12} - \left(\frac{\sigma}{r}\right)^{6}\right)$$

All particles have a fixed value of the dipole moment which contributes to the interaction energy of each i-th and k-th particle is given by expression:

$$_{pp} = \frac{1}{r^3}(\vec{p}_k * \vec{p}_j) - \frac{3}{r^5}(\vec{p}_k * \vec{r})(\vec{p}_j * \vec{r})$$

Where $\vec{p}_k$ and $\vec{p}_j$ - dipole moments of the i-th and k-th particle respectively, $\vec{r} = \vec{r}_k - \vec{r}_j$, $r = |\vec{r}|$. Modeling was performed in LJ units: $\sigma = 1$, $\varepsilon = 1$. The simulations were performed for two-dimensional systems with particle concentration $n = 0.8$. The initial velocity distribution was specified in accordance with the Maxwell distribution. The dipole moments of the system in the initial state are oriented randomly. Temperature of the system was maintained at T = 1, for this Brendstone thermostat was used. After setting the initial state of the system modeling was performed during $t = 100$ in LD units in order to obtain the equilibrium state. Then simulation of system was performed during $t = 500$ in LD units in order to obtain experimental data.

In the second phase of modeling at each simulation step calculation of average value of dipole moment $\langle\vec{p}_k(t)\rangle$, average speed of rotary motion of the particles $\Omega(t) = \langle\omega(t)\rangle$ and the quantities $P(t) = (\langle\vec{p}_k(t)\vec{p}_k(0)\rangle$ was performed (the brackets $\langle\ \ \rangle$ denote averaging over all systems particles).

The simulations were performed for different PDM values varying in the range from 0.1 to 2.5 in increments of 0.1 in LD units. The MD simulations are performed using the LAMMPS simulation package [9]. Visualization of the resulting structures was performed in the package VMD [10].

3. Results

Figure 1 shows snapshots of the systems structures obtained for different values of the PDM. System has a rather homogeneous structure at low values of PDM (Fig. 1a). Movement of particles leads to the appearance of cavities in systems structure. However they have relatively small size and very short lifetime. Increasing the value of PDM leads to changes in systems structure: lifetime of cavities increases significantly and their sizes increased up to a few interatomic distances (figure 1b). It is also possible to observe the formation of dipole chains, but they have very small length and lifetime. At some critical value of PDM occurs systems polymerization: stable long dipole chains are formed in system (figure3b). Also, there are large long-lived cavities in system.

Figure 2 shows the autocorrelation function (ACF) P (t) for systems with different values of PDM. It should be noted that form of ACF P (t) varies continuously when PDM changed in $p \lesssim 2.1$ region. However form of ACF P(t) varies discontinuously and oscillations P (t) become highly correlated when p ~ 2.1 which corresponds to system polymerization.

Journal of Physics: Conference Series **486** (2014) 012031 doi:10.1088/1742-6596/486/1/012031

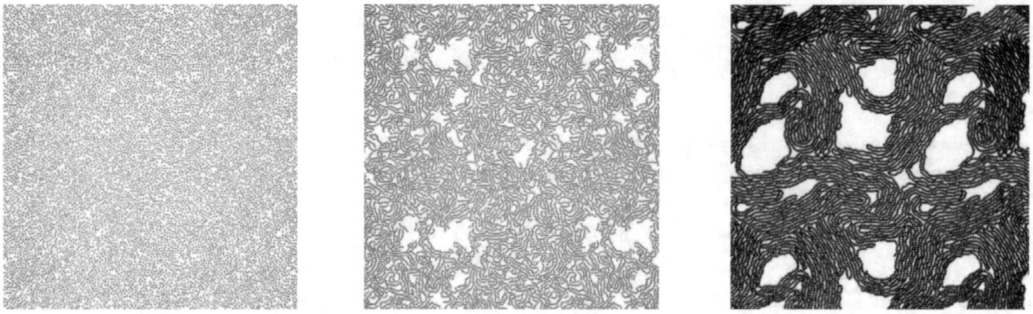

(a) (b) (c)

Figure 1. Snapshots of the system structure for different values of PDM: a – $p = 0.5$; б – $p = 1.5$; г – $p = 2.5$.

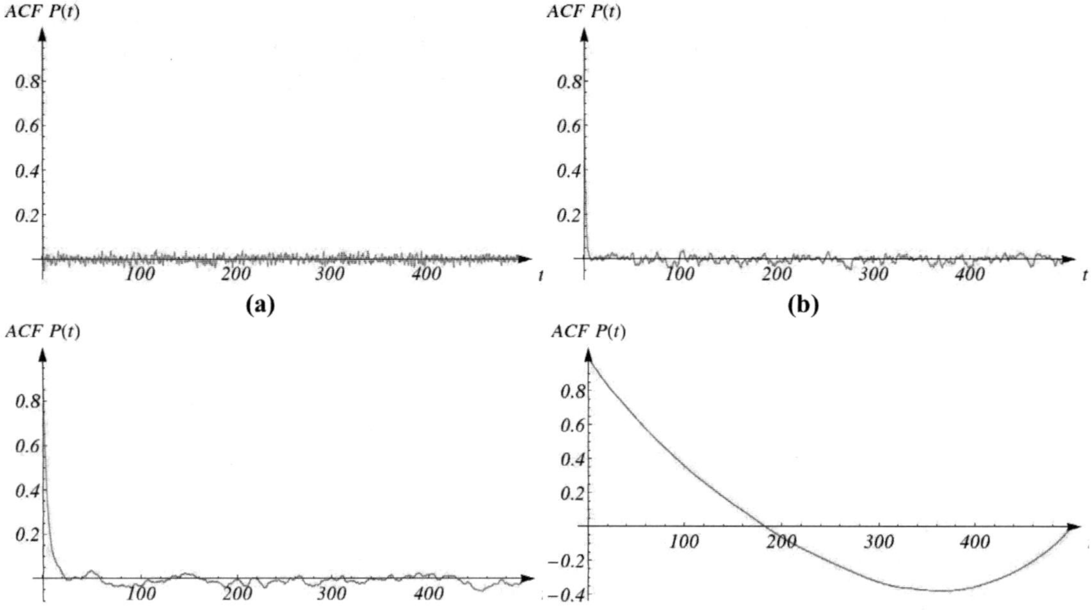

(c) (d)

Figure 2. Autocorrelation functions of the P(t) for systems with different values of PDM: a – $p = 0.5$; b – $p = 1.5$; c – $p = 2.0$; d – $p = 2.5$

In figures 3-4 shows power spectrum of Ω (t) and P (t), respectively, for systems with different values of PDM. It can be seen that increasing of PDM values leads to a shift of the Ω(t) power spectrum into high-frequency region, while the P(t) power spectrum is shifted to lower frequencies.

Conclusions

Thus it was shown that the increase of PDM values leads to formation of long-lived cavities whose sizes are comparable with several interatomic distances. Also increasing of PDM values leads to a shift of the Ω(t) power spectrum into high-frequency region, while the P(t) power spectrum is shifted to lower frequencies. It has been shown that there is a critical ratio between the temperature and PDM values at which system is polymerized and ACF P(t) undergoes discontinuous change due to the fact that oscillations of the dipole moments become strongly correlated.

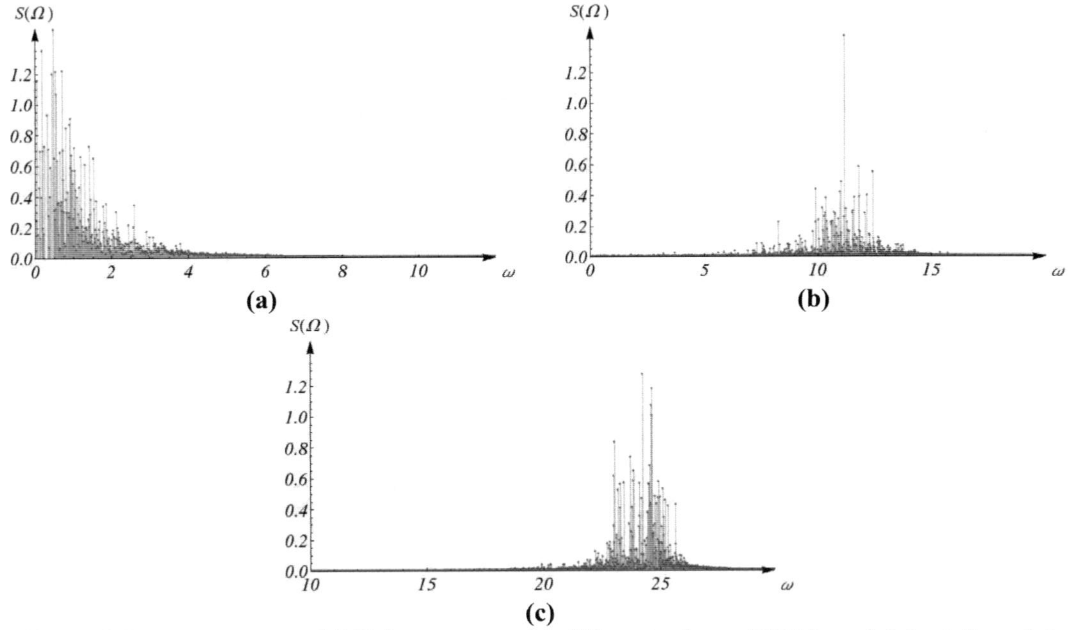

Figure 3. Power spectrum of Ω(t) for systems with different values of PDM: a - 0.5; b - 1.5; c - 2.5

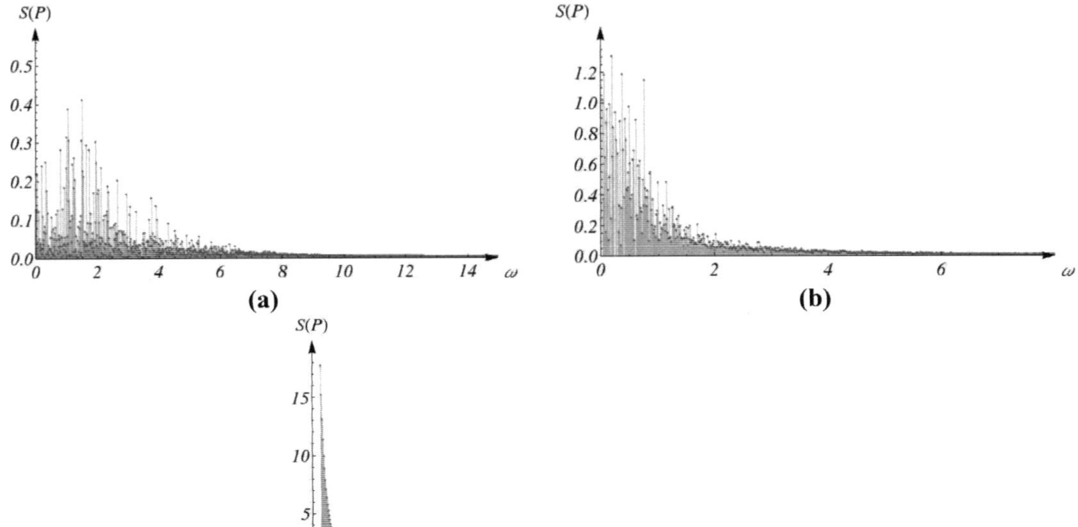

Figure 4. Power spectrum of P(t) for systems with different values of PDM: a - 0.5; b - 1.5; c - 2.5

Acknowledgments

The authors are grateful to Professor. N.F. Bunkin for fruitful discussion of the results. This work was supported by grants of the Ministry of Education and Science of Russian Federation (№ 14.B37.21.0898, № 14.B37.21.1282) and of the Russian Foundation for Basic Research (RFBR 12-08-31104; RFBR 12-08-33112).

References

[1] Aliev I N, Yurchenko S O 2010 Evolution of perturbations of a charged interface between immiscible inviscid fluids in the interelectrode gap *Fluid Dynamics* **45**(5) 817-826

[2] Aliev I N, Yurchenko S O, Nazarova E V 2007 On the problem of instability of the boundary of two media of finite thickness *J. Eng. Phys. Thermophys.* **80**(6) 1199-120

[3] Aliev I N, Yurchenko S O 2009 Nonlinear waves propagating over a conducting ideal fluid surface in an electric field *Fluid Dynamics* **44**(5) 748-758

[4] Bunkin N F, Yurchenko S O, Suyazov N V, Shkirin A V 2012 Structure of the nanobubble clusters of dissolved air in liquid media *J. Biol. Phys.* **38**(1) 121-152

[5] Milischuk A A, Ladanyi B M 2011 Structure and dynamics of water confined in silica nanopores *J. Chem. Phys.* **135** 174709

[6] Kolev V, Ivanova A, Madjarova G, Aserin A, Garti N 2012 Molecular dynamics approach to water structure of HII mesophase of monoolein *J. Chem. Phys.* **136** 074509

[7] Dickey A N, Stevens M J 2012 Site-dipole field and vortices in confined water *Phys. Rev. E.* **86** 051601

[8] Tan M-L, Fischer J T, Chandra A, Brooks B R, Ichiye T 2003 A temperature of maximum density in soft sticky dipole water *Chem. Phys. Lett.* **376** 646-652

[9] LAMMPS Molecular Dynamics Simulator http://lammps.sandia.gov/

[10] *Visual Molecular Dynamics,* http://www.ks.uiuc.edu/Research/vmd/

[11] Bunkin N, Yurchenko S, Suyazov N, Shkirin A 2012 Structure of the nanobubble clusters of dissolved air in liquid media *J. Biologic. Phys.* **38** 121–152

[12] Toukan K and Rahman A 1985 Molecular-dynamics study of atomic motions in water *Phys. Rev. B* **31** 2643–2648

[13] Bertrand G 2002 A reappraisal of what we have learnt during three decades of computer simulations on water *J. Mol. Liquids* **101** 219–260

Plasma resonant terahertz photomixers based on double graphene layer structures

Maxim Ryzhii[1], Michael S Shur[2], Vladimir Mitin[3], Akira Satou[4], Victor Ryzhii[4,5] and Taiichi Otsuji[4]

[1]Department of Computer Science and Engineering, University of Aizu, Japan
[2]Department of Electrical, Electronics, and Systems Engineering, Rensselaer Polytechnic Institute, Troy, NY, USA
[3]Department of Electrical Engineering, University at Buffalo, Buffalo, NY, USA
[4]Research Institute for Electrical Communication, Tohoku University, Sendai, Japan
[5]Research and Educational Center "Photonics and Infrared Technology", Bauman Moscow State Technical University, Russia

E-mail: m-ryzhii@u-aizu.ac.jp

Abstract. We propose terahertz (THz) photomixers based on double graphene layer (DGL) structures, utilizing the interband absorption of modulated optical radiation, tunneling or thermionic inter-GL transitions, and resonant excitation of plasma oscillations. Using the developed device model, we substantiate the operation of the photomixers and calculate their characteristics. We demonstrate that the output frequency-dependent power of THz radiation exhibits pronounced resonant peaks at the plasmonic resonant frequencies. The proposed THz photomixer can surpass the pertinent devices based on the standard heterostructures.

1. Introduction

The energy spectrum of graphene layers (GLs) provides the interband photogeneration of electrons and holes by electromagnetic radiation from the terahertz to ultraviolet range. In recently fabricated and analyzed DGL structures [1],[2], the propagating and standing plasma waves can exhibit nontrivial properties. In particular, in the DGL structures the waves, in which the electron and hole densities oscillate in the same phases, are similar to those in the gated structures and characterized by the linear plasma wave dispersion relation. This is because in these structures each GL plays the role of a highly conducting gate for the other GL [3],[4].

In this paper, we propose THz photomixers comprising the DGL structures irradiated by two coherent optical beams produced by lasers with close photon frequencies, Ω_1 and Ω_2 (so that the modulation frequency $\omega = \Omega_2 - \Omega_1$) or by ultra-short optical pulses for the generation of THz radiation (see also [5]). The photomixer operation is associated with the ac photocurrent between GLs at the difference frequency ω. The radiation-induced photocurrent between GLs results in the ac terminal current. The latter feeds an antenna, which emits the radiation with the frequency ω. When ω approaches to the frequencies of the plasma oscillation modes, the excitation of plasma oscillations in the DGL structures occurs, the ac inter-GL photocurrent and the power of output radiation can be resonantly large.

Content from this work may be used under the terms of the Creative Commons Attribution 3.0 licence. Any further distribution of this work must maintain attribution to the author(s) and the title of the work, journal citation and DOI.
Published under licence by IOP Publishing Ltd

Journal of Physics: Conference Series **486** (2014) 012032 doi:10.1088/1742-6596/486/1/012032

2. Device structures and model

We consider DGL photomixers with undoped GLs and the inter-GL barrier made of a material like Tungsten Disulfide (WS_2), which forms a relatively low barrier, Δ_C, for electrons (and rather high one for holes) or of a very thin layer with sufficient tunneling transparency for the photoexcited electrons. The device structures under consideration and their band diagrams at the applied bias voltage V_0 are shown in figures 1 and 2, respectively.

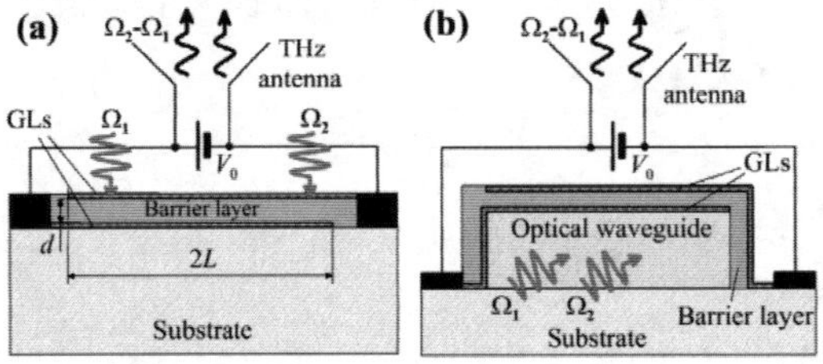

Figure 1. Schematic view of the DGL photomixer structures with vertical optical input (a) and the DGL structures integrated with waveguide (b).

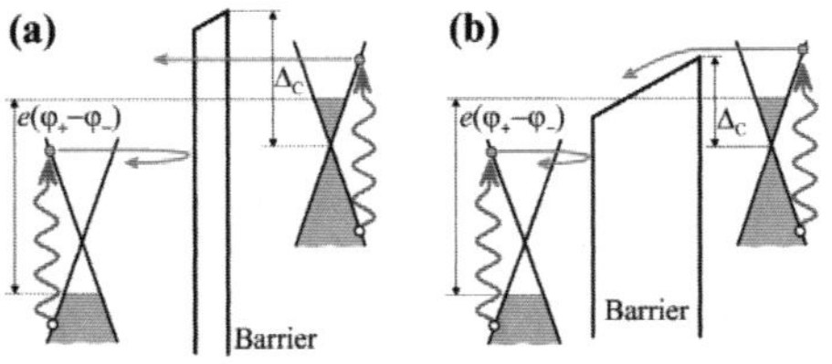

Figure 2. Energy band diagram of the DGL photomixers with tunneling (a) and thermionic (b) barriers. Wavy arrows correspond to photogeneration of electron-hole pairs by photons with frequencies $\Omega_1 \approx \Omega_2$. Smooth arrows denote inter-GL transitions. Quantity $e(\varphi_+ - \varphi_-)$ is local potential difference between GLs.

It is assumed that each GL is supplied with a side contact. The bias voltage, V_0, induces the extra electrons in upper GL and extra holes in lower GL. The intensity of the incoming optical radiation I comprises the ac component (the beating intensity) $\delta I \exp(-i\omega t)$.

We describe the dynamics of the electron-hole plasma perturbed by the modulated optical radiation using the linearized hydrodynamic equations coupled with the Poisson equation in the gradual channel approximation. The set of the equations of the model is reduced to the following coupled equations for the ac potentials $\delta\varphi_+$ and $\delta\varphi_-$ of the upper and lower GLs:

$$\frac{d^2\delta\varphi_+}{dx^2} + \frac{\omega(\omega+i\nu)}{s^2}\left(\delta\varphi_+ - \delta\varphi_-\right) = -\delta j(\nu - i\omega)\frac{4\pi d}{ks^2}, \quad (1)$$

$$\frac{d^2\delta\varphi_-}{dx^2} + \frac{\omega(\omega+i\nu)}{s^2}\left(\delta\varphi_- - \delta\varphi_+\right) = \delta j(\nu - i\omega)\frac{4\pi d}{ks^2}. \quad (2)$$

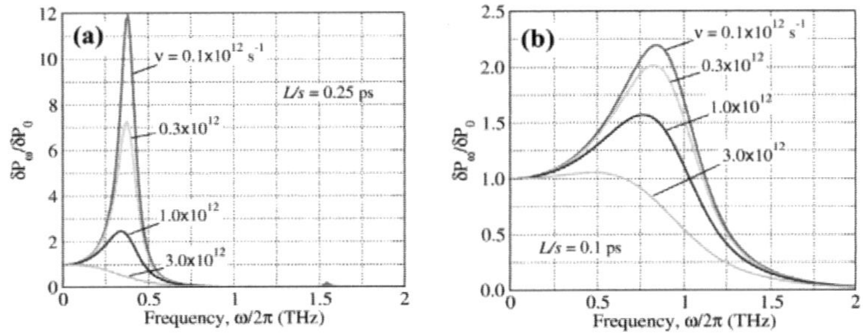

Figure 3. Normalized output power $\delta P_\omega/\delta P_0$ versus frequency for different values of electron and hole collision frequency ν and for L/s = 0.25 ps (a), and for L/s = 0.1 ps (b).

Here ν is the collision frequency of electrons and holes in GLs with impurities and acoustic phonons, δj is the density of the inter-GL ac current, d is the barrier thickness, k is the dielectric constant, and s is the characteristic velocity of plasma waves in GLs. The terms in the right-hand sides of equations (1) and (2) are associated with the ac inter-GL current. The plasma wave velocity s is determined by the net dc electron and hole density, i.e., by the applied voltage: $s \propto (V_0 d)^{1/4}$. The boundary conditions are presented by the following:

$$\delta\varphi_\pm\Big|_{x=\pm L} = \pm\frac{\delta\varphi_A}{2}, \quad \frac{d\delta\varphi_\pm}{dx}\Big|_{x=\mp L} = 0\,, \qquad (3)$$

where $\delta\varphi_A$ is the drop of the ac potential across the antenna, which is proportional to its radiation resistance r_A and the terminal ac current δJ, and $2L$ is the lateral size of GLs (see figure 1).

3. Results

Solving equations (1) and (2) with boundary conditions (3), we found the spatial distribution of the ac potentials in GLs, the net terminal ac current, and the output THz power. The latter is described by the following equations:

$$\delta P_\omega = \frac{\delta P_0}{(1+\omega^2\tau_e^2)}\left|Q_\omega\left(\cot Q_\omega - Q_\omega\right) - i\omega\tau_A\right|^{-2}, \qquad (4)$$

$$Q_\omega = \frac{\pi\sqrt{\omega(\omega+i\nu)}}{2\omega_p}\,. \qquad (5)$$

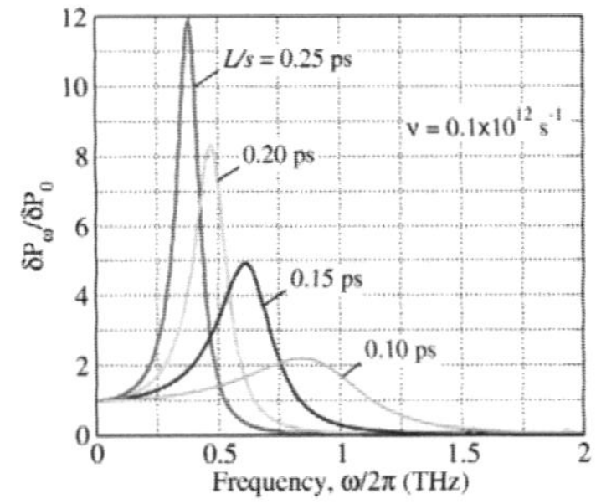

Figure 4. Normalized output power $\delta P_\omega/\delta P_0$ versus frequency for different values of parameters L/s corresponding to the plasma frequencies 0.77 −1.77 THz

Here $\delta P_0 = r_A R_0^2 \delta P_\Omega^2/2$ is the output power at small ω, R_0 is low frequency responsivity proportional to the probability, w_{esc}, of the photoexcited electron escape from one GL to another, δP_Ω is the amplitude of incident optical power modulation, τ_e is the time of the inter-GL and intra-GL relaxation of photoexcited electrons,

Journal of Physics: Conference Series **486** (2014) 012032 doi:10.1088/1742-6596/486/1/012032

$\tau_A \propto r_A$ is the antenna recharging time, and $\omega_p = \pi s/(2\sqrt{2L})$ is the characteristic plasma frequency.

The frequency dependencies of the output THz power calculated using equations (4) and (5) are demonstrated in figures 3-5 for different structural parameters. The obtained characteristics exhibit pronounced resonant peaks associated with the excitation of plasma oscillations. As follows from figures 3-5, the peak values of the output THz power at the elevated values of plasma oscillations quality factor ω_p/ν substantially exceed the latter at low frequencies.

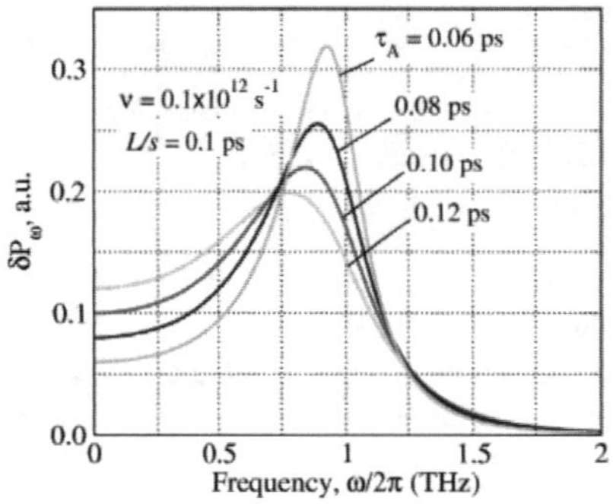

Figure 5. Frequency dependences of output power δP_ω for different values of recharging time τ_A.

Acknowledgements
This work was supported by the Japan Society for Promotion of Science, PIRE TeraNano Program, NSF, USA, and by the Army Research Laboratory, USA.

References

[1] Liu M et al 2011 A graphene-based broadband optical modulator *Nature* **474** 64
[2] Britnel L et al 2012 Field-effect tunneling transistor based on vertical graphene heterostructures *Science* **335** 947
[3] Ryzhii V et al 2012 Double graphene-layer plasma resonances terahertz detector *J. Phys. D: Appl. Phys.* **45** 302001
[4] Ryzhii V et al 2012 Effect of plasma resonances on dynamic characteristics of double graphene-layer optical modulator *J. Appl. Phys.* **112** 104507
[5] Ryzhii V et al 2013 Terahertz photomixing using plasma resonances in double-graphene layer structures *J. Appl. Phys.* **113** 174506-1

Journal of Physics: Conference Series **486** (2014) 012033 doi:10.1088/1742-6596/486/1/012033

THz photoluminescence from n-GaN layers at CW interband excitation

Alexander V Andrianov[1], Alexey O Zakhar'in[1], Alexander V Bobylev[1] and Zhe C Feng[2]

[1] A. F. Ioffe Physical Technical Institute, 194021, St. Petersburg, Russia
[2] Institute of Material Research and Engineering, 119260, Singapore

E-mail: Alex.Andrianov@mail.ioffe.ru

Abstract. We report on experimental observation and studies of terahertz emission from n-doped GaN films under continuous wave interband photoexcitation at low temperatures. Properties of the terahertz emission testify that the emission is caused by intracenter optical transitions in shallow donors of GaN. The THz intracenter optical transitions are initiated by the interband photoexcitation.

1. Introduction and background

It is known for a long time that terahertz (THz) radiation can occur under optical transitions of electrons between the energy levels of shallow impurities in semiconductors [1],[2]. The THz radiative transitions arise from the energy relaxation of nonequilibrium carriers, created in allowed bands, for example, as a result of impact ionization of impurity centers [1],[2]. Recently we showed that the impurity related THz emission in doped semiconductors can be achieved at the interband optical excitation at low temperatures [3],[4]. These results open an intriguing possibility to achieve the impurity related THz photoluminescence from wide band gap semiconductors and, in particular, from GaN, where intracenter THz electroluminescence was observed only at quite high electric fields [5].

Here we report on THz emission from n-GaN(Si) epitaxial layers under continuous-wave band-to-band photoexcitation.

2. Experimental details

The studied GaN epitaxial layers were grown by MOCVD method on a 450 μm thick sapphire (0001) substrate and had thickness of about 1.8 – 2 μm. The GaN layers were doped with Si during growth process. Carrier concentration and their mobility at 300 K were 4×10^{17} cm^{-3} and 190 $cm^2/V\times s$, consequently. A CW He-Cd laser (λ=325 nm, peak power of 8 mW) was used for interband photoexcitation of the samples. The laser beam was focused on studied GaN epitaxial layers through sapphire substrate. The THz emission from the sample was analyzed in collinear geometry with the excitation beam. Spectral measurements were done using a step-scan Fourier spectrometer for the 5-350 cm^{-1} spectral range, described elsewhere [6]. The THz radiation was detected by liquid helium cooled Si bolometer with using lock-in technique.

Content from this work may be used under the terms of the Creative Commons Attribution 3.0 licence. Any further distribution of this work must maintain attribution to the author(s) and the title of the work, journal citation and DOI.
Published under licence by IOP Publishing Ltd

3. Results

The THz photoluminescence (PL) was observed in the temperature range from 5 to ~140 K. The figure 1 shows typical spectrum of THz emission form n-GaN(Si) layers. The THz PL spectrum demonstrates pronounced maxima at ~23 and 26 meV (~5.57 and 6.29 THz), which can be assigned to the 2P -1S intracenter optical transitions in shallow donor centers related to Si and O impurities, respectively, in GaN.

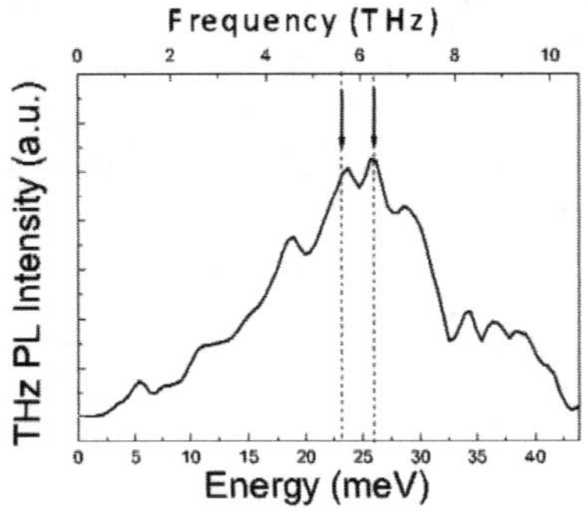

Figure 1. Typical spectrum of THz PL from n-GaN(Si) layers at T=5 K and excitation density of ~0.6 W/cm^2. The arrows points the energies of 23 and 25 meV known for 2P-1S optical transitions in donors related to Si and O impurities, respectively [5],[7].

The emission intensity decreases with increase in temperature. The Arrhenius plot of the THz PL intensity temperature dependence demonstrates a temperature quenching energy of the emission of the order of 33.5 meV that is consistent with the binding energy of shallow donors in GaN. The dependence of THz emission intensity versus the excitation density shows clear sublinear character. This dependence can be well approximated by a square root law.

The properties of the discovered THz PL allow it to be attributed to the emission, which occurs under the capture of nonequilibrium electrons to ionized donor centers. The photoactive ionized donors can be created in n-doped material at low-temperatures as a result of recombination of free holes with electrons localized on donor centers (h-D^0 recombination). In fact the near-band-edge PL spectrum of the studied GaN layers at 5 K demonstrates the distinct line at 3.466 eV, which corresponds to the h-D^0 recombination process.

The discovered THz PL can find applications in THz technologies.

Acknowledgements

This work was supported by the Russian Foundation of Basic Research (# 11-02-00230 and # 12-02-00455a projects).

References

[1] Koenig S H, Brown R D 1960 *Phys. Rev. Lett.* **4** 170
[2] Melngailis I et al 1969 *Phys. Rev. Lett.* **23** 1111
[3] Andrianov A V et al 2010 *JETP Lett.* **91** 96
[4] Zakhar'in A O et al 2010 *Appl. Phys. Lett.* **96** 211118
[5] Shalygin V A et al 2009 *J. Appl. Phys.* **106** 123523
[6] Andrianov A V et al 2004 *JETP Lett.* **79** 448
[7] Moore W J et al 2001 *Appl. Phys. Lett.* **79** 2570

Journal of Physics: Conference Series **486** (2014) 012034 doi:10.1088/1742-6596/486/1/012034

An approach for automatic construction of the wavelet-domain de-noising procedure for THz pulsed spectroscopy signal processing

Kirill I Zaytsev, Nikita V Chernomyrdin, Alexey V Gorevoy, Nikolay E Trofimov, Irina N Fokina, Valentin I Alekhnovich, Valeriy E Karasik and Stanislav O Yurchenko

Bauman Moscow State Technical University, 2nd Baumanskaya str., 5, Moscow, 105005, Russia

E-mail:kirzay@gmail.com

Abstract. De-noising of terahertz pulsed spectroscopy (TPS) signals is an essential problem, since a noise in the TPS data samples makes correct reconstruction of sample spectral dielectric properties and internal structure challenging. It is especially important for the spectral regions where detector sensitivity is typically low. A lot of effective techniques for $1D$ and $2D$ signal de-noising based on the signal processing in wavelet-domain have been developed in recent times. The present work demonstrates the ability to perform effective de-noising of pulsed spectroscopy signals using the algorithm of the Fast Wavelet Transform (FWT). The results of optimal wavelet basis selection and the results of adaptive wavelet-domain filter selection are reported. The performance of the wavelet-domain de-noising algorithm implementation is also discussed. A technique for automatic construction of the wavelet-domain de-noising procedure is offered.

1. Introduction and background

Terahertz (THz) pulsed spectroscopy [1] (TPS) is an essential and effective tool for measuring the THz dielectric properties of a sample, including spectral dependencies of complex dielectric permittivity of a homogeneous sample [2]-[4] and permittivity and conductivity profiles characterizing the internal structure of an object [5]-[7]. The last technique, which has been described in the paper [7] for the first time, is called THz tomography or T-ray tomography.

In TPS the sample of interest is radiated with a short pulse of THz radiation, which form is close to single electrical field oscillation. After transmission through the sample or reflection from the sample surface, the electric field is detected with very high time resolution: it reaches $50.0\,fs$ for the TPS system utilized in the present work. Then the fast Fourier transform (FFT) algorithm is applied to analyze spectrums of the detected THz waveforms and reconstruct the THz spectral dielectric properties of the sample.

Regardless the type of the data we want to extract from the detected THz waveforms, the accuracy of the result is strongly correlated with the signal-to-noise ratio. The time-domain signal-to-noise ratio SNR_t is constant for all of time-domain delays t, but the signal-to-noise ratio in the Fourier-domain SNR_f significantly depends on the frequency of the signal harmonics [8]. The power of the detected signal is high for the medium frequency range (for our time-domain system (TDS) [5] system it corresponds to the range from 0.3 to $1.5\,THz$), and the SNR_f is rather high. Depending on the THz

Content from this work may be used under the terms of the Creative Commons Attribution 3.0 licence. Any further distribution of this work must maintain attribution to the author(s) and the title of the work, journal citation and DOI.
Published under licence by IOP Publishing Ltd

pulse source and detector types, and on the number of averaged waveforms, the SNR_f can be $100.0 \ldots 5000.0\, arb.units$. Otherwise, for the low- and high-frequency ranges (for utilized TDS system these ranges are $0.1 \ldots 0.3\, THz$ and $1.5 \ldots 3.0\, THz$, respectively) the SNR_f is smaller than $10.0 \ldots 100.0\, arb.units$. Therefore the results of the sample spectroscopic analysis have high accuracy in the region of spectrum corresponding to 0.3 to $1.5\, THz$ and lower accuracy in the $0.1 \ldots 0.3\, THz$ and $1.5 \ldots 3.0\, THz$ regions.

The purpose of the present work is to develop an effective algorithm for THz waveform de-noising, which can help to study the sample properties with maximal accuracy in the entire range of TPS spectral sensitivity. A lot of recent works are dedicated to the wavelet-domain processing of $1D$ and $2D$ data [9],[10] and the wavelet de-noising of TPS signals [11]-[13]. The peculiarity of the wavelet transform consists in the existence of a large amount of wavelet-bases and wavelet spectrum processing techniques which can be used for noise suppression. The selection of optimal wavelet basis and the selection of the most effective noise suppressing technique are important problems, which need to be solved during the development of any de-noising algorithm. An approach to the solution of these problems will be described in the present work.

The wavelet analysis is a generalization of the spectral analysis techniques [9]. The typical and the most common technique for the signal spectral analysis is the Fourier analysis, which includes different methods for the direct and the inverse Fourier transform implementation and signal processing methods. The Fourier transform can be described as a projection of a signal $s(t)$ on the basis of periodic functions $e^{-2\pi\nu_t t}$ with different frequencies ν_t and infinite width:

$$\tilde{s}(\nu_t) = \frac{1}{\sqrt{2\pi}} \int_{-\infty}^{+\infty} s(t)\, e^{-2\pi\nu_t t} dt. \tag{1}$$

The main drawback of Fourier analysis is poor time-domain localization. For instance, if we consider that the useful part of the signal is presented only in the time interval t_1, and the whole time interval of analysis is t_2, then the SNR_f can be described as follows:

$$SNR_f \sim \frac{t_1}{t_2}. \tag{2}$$

Here we use an assumption of white noise, so the noise power in the frequency-domain is constant and proportional to t_2. Obviously, the more local information we deal with, the less SNR_f we could obtain.

To overcome this drawback, we could use the windowed Fourier transform, but it is redundant and difficult to invert. A better solution is to use wavelets. In contrast to Fourier analysis, the wavelet-domain signal processing provides the time-domain resolution, since the wavelet-transform is equal to the projection of a signal on the basis of local functions (wavelets). The discrete wavelet transform (DWT) can be described as the following series representation of a function $s(t)$ [9]:

$$s(t) = \sum_{a=a_0}^{a_N} \sum_{b=0}^{b_M} C_{a,b} \psi_{a,b}(t), \tag{3}$$

where the coefficients of the wavelet-spectrum $C_{a,b}$ are defined with an equation:

$$C_{a,b} = \langle s(t)\psi_{a,b}(t) \rangle = \int_{-\infty}^{+\infty} s(t)\psi_{a,b}(t) dt \tag{4}$$

Equation (4) describes the projection of a signal $s(t)$ on the basis of wavelet kernels $\psi_{a,b}(t)$ with different kernel scales a and various kernel offsets b.

The form of wavelet kernels $\psi_{a,b}(t)$ can vary depending on the type of the basis we use for signal analysis. In the DWT the number of kernels is strongly restricted. All of the kernels are orthogonal to each other:

$$\langle \psi_{a,b}(t)\psi^*_{a',b'}(t)\rangle = \delta_{a,a'}\delta_{b,b'}, \tag{5}$$

where $\delta_{i,j}$ is a Kronecker symbol. Also, kernels should satisfy the following normalization condition:

$$\int_{-\infty}^{+\infty} \psi_{a,b}(t)dt = 0; \quad \left\|\psi_{a,b}(t)\right\|^2 = \int_{-\infty}^{+\infty} \left|\psi_{a,b}(t)\right|^2 dt < \infty. \tag{6}$$

Note that this condition shows that the projection of a function $s(t)$ on wavelet kernels $\psi_{a,b}(t)$ is equivalent to the band-pass filtration of a signal in the Fourier-domain with the frequency of maximal filter transparency equal to $1/a$. For the DWT and the fast wavelet transform (FWT) algorithms the relationship between scales is presented by the following equation: $a_{i+1} = 2a_i$. A solution for discrete step of wavelets can be derived from the basis orthogonality condition. More details about the DFT and the FWT implementation can be found in [9].

The FWT should use a limited set of wavelet transform bases due to the specific restrictions [9],[10]. Such families of wavelet bases as Haar (*'haar'*), Daubechies (*'db'*), Symlets (*'sym'*), Coiflets (*'coif'*), Biorthogonal (*'bior'*), Reversed Biorthogonal (*'rbio'*), Meyer's Discrete Wavelets (*'dmay'*) are often used for the FWT and presented in all wavelet processing toolboxes. All of these families can contain one wavelet basis, for instance, *'haar'* basis or *'dmay'* basis, or several bases, for example, *'db1'*, *'db2'*, *'db3'*,..., *'coif1'*, *'coif2'*, *'coif3'*, ..., *'bior1.1'*, *'bior1.3'*, *'bior1.5'*, *'bior2.2'*, *'bior2.4'*, *'bior2.6'*, etc. The order of the wavelet basis in the family correlates with the sharpness of the locality of the wavelets in time-domain frequency domain.

The wavelet-domain signal-to-noise ratio SNR_w is proportional to the following constant:

$$SNR_w \sim \begin{cases} 1, & if \quad t_1 \geq t_3, \\ \dfrac{t_1}{t_3} & if \quad t_1 < t_3, \end{cases} \tag{7}$$

Here t_1 is defined the same way as in the equation (2), and t_3 is the size of the wavelet kernel. We use the same assumption of white noise. One can compare the equations (2) and (7). Since the width of the wavelet kernel t_3 is smaller than the whole time interval of analysis t_2, we can conclude that the wavelet-domain signal-to-noise ratio SNR_w is higher than the Fourier-domain signal-to-noise ratio SNR_w. This condition leads to benefits of the wavelet-domain de-noising approach in comparison to the de-noising in the Fourier-domain.

The next part of the paper presents the results of the wavelet-domain de-noising procedure implementation for processing the typical data samples acquired with TDS systems.

2. Results

2.1. Selection of the optimal wavelet basis for TDS system signals representation. The problem of the optimal wavelet basis selection is an essential one, since the usage of the wavelet basis, incompatible with the TDS signals can cause non-linear signal distortions.

An approach for the optimal wavelet basis selection for TDS signals processing has been presented in the work [11]. The proposed technique utilizes several criteria to evaluate compatibilty of signal $s(t)$ and wavelet kernels $\psi_{a,b}(t)$, including:

- The Wavelet Basis Efficiency Index $WBEI(s(t), \psi_{a,b}(t))$;
- The Pulse Spectral Relative Entropy Criterion $PSRE(s(t), \psi_{a,b}(t))$;
- The Pulse Spectral Cumulative Error Criterion $PSCE(s(t), \psi_{a,b}(t))$.

The results of listed criteria calculation for the TDS system signal, presented in the paper [11], show that the wavelet bases *'bior3.3'* is the most optimal for analysis of the considered TDS signals. In the present work we consider the TDS system based on LT-GaAs photoconductive antenna source of THz pulses and ZnTe electrooptical detector of THz field. Our system utilizes ultrashort optical pulses with 80.0 fs duration. Obviously, the optimal wavelet basis for the signals of our TDS system can differ from the described one.

We use the following generalized criterion for the selection of several wavelet bases, which are the most optimal for the present TDS system signal processing:

$$\begin{aligned} G\left(s(t), \psi_{a,b}(t)\right) = \alpha \cdot WBEI_N\left(s(t), \psi_{a,b}(t)\right) \\ + \beta \cdot PSRE_N\left(s(t), \psi_{a,b}(t)\right) \\ + \gamma \cdot PSCE_N\left(s(t), \psi_{a,b}(t)\right), \end{aligned} \tag{8}$$

where α, β, and γ are coefficients, regulating the impact of each described criterion, and $WBEI_N\left(s(t), \psi_{a,b}(t)\right)$, $PSRE_N\left(s(t), \psi_{a,b}(t)\right)$, and $PSCE_N\left(s(t), \psi_{a,b}(t)\right)$ are the normalized criteria for target wavelet basis:

$$\begin{aligned} WBEI_N\left(s(t), \psi_{a,b}(t)\right) &= \frac{WBEI\left(s(t), \psi_{a,b}(t)\right)}{\max_{\psi_{a,b}(t)}\left\{WBEI\left(s(t), \psi_{a,b}(t)\right)\right\}}, \\ PSRE_N\left(s(t), \psi_{a,b}(t)\right) &= \frac{PSRE\left(s(t), \psi_{a,b}(t)\right)}{\max_{\psi_{a,b}(t)}\left\{PSRE\left(s(t), \psi_{a,b}(t)\right)\right\}}, \\ PSCE_N\left(s(t), \psi_{a,b}(t)\right) &= \frac{WBEI\left(s(t), \psi_{a,b}(t)\right)}{\max_{\psi_{a,b}(t)}\left\{PSRE\left(s(t), \psi_{a,b}(t)\right)\right\}}, \end{aligned} \tag{9}$$

Here we use the following relation between the criterion impacts: $\alpha = \beta = \gamma$. After calculation of the generalized criterion (8) for different wavelet bases, we have selected the following set of optimal wavelet bases for our TDS system.

Table 1. Optimal wavelet-bases for TDS signal processing

Wavelet Family Name	**Daubechies** ('*db*')			**Symlets**('*sym*')	**Coiflets** ('*coif*')	
Wavelet Basis Name	*'db5'*	*'db6'*	*'db7'*	*'sym8'*	*'coif3*	*'coif5'*

The test signal which has been used for the optimal wavelet basis selection is presented in the figure 1. All the bases of the described set shows high value of the compatibility with the test signal $s(t)$ and can provide high de-noising accuracy. We compare this set of wavelets again on the step of the de-noising technique implementation.

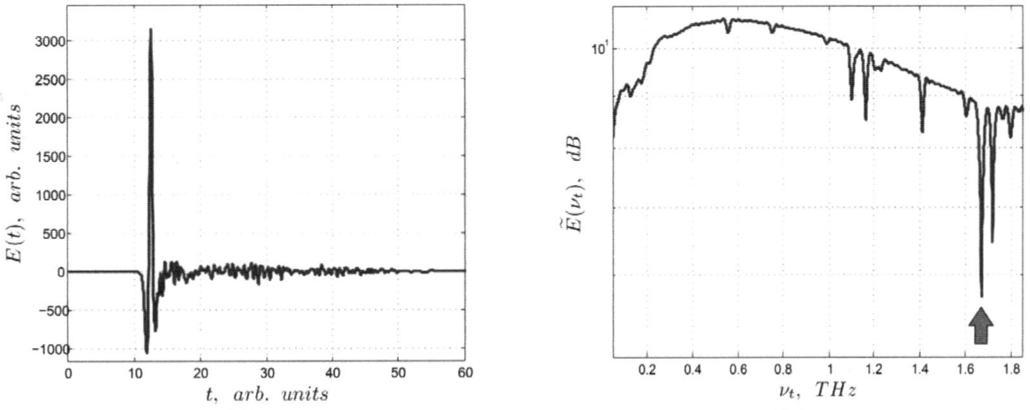

Figure 1. The test TDS signal s(t) utilized for the optimal wavelet basis selection: (a) – time-domain representation of the signal, (b) – Fourier-domain representation of the signal. The pointer in 1(b) marks the water vapour absorption line which is used during the comparison of the different filtration techniques.

2.2. Selection of the wavelet domain de-noising method. A lot of wavelet-domain de-noising techniques exist. Wavelet de-noising procedure is associated with the thresholding of wavelet-domain spectrum coefficients $C_{a,b}$ representing a signal $s(t)$:

$$C_{a,b} = \begin{cases} C_{a,b}, & if \quad C_{a,b} > T, \\ 0, & if \quad C_{a,b} < T, \end{cases} \tag{10}$$

where T is the certain threshold value. Several common techniques for calculation of T are described in [9],[10]. In the present paper we consider and compare the following thresholding procedures [9]:

- Minimax thresholding (*'MiniMax'*);
- Heuristic threshold (*'Heursure'*);
- $T = \sqrt{2 * log(l)}$ threshold (*'SqTwoLog'*),
  where l – is the length of the certain wavelet level a' array.

Since the extraction of useful information from registered signal (determination of the sample spectral properties, material parameters) involves the processing of TDS signal in Fourier-domain, the accuracy of spectral characteristics reconstruction depends directly on the accuracy of TDS single signal determination in Fourier-domain $s(t)$. The technique of wavelet-domain de-noising should suppress the noise in the time-domain and in the Fourier-domain, respectively, but it also should not cause non-linear distortions to the Fourier-domain TDS. The criterion for the filtering procedure comparison should be based on the Fourier-domain data analysis.

To compare thresholding techniques the white Gaussian noise with certain standard deviation σ_N was added to the test signal $s(t)$. The noised signal and its spectrum are presented in the figure 2.

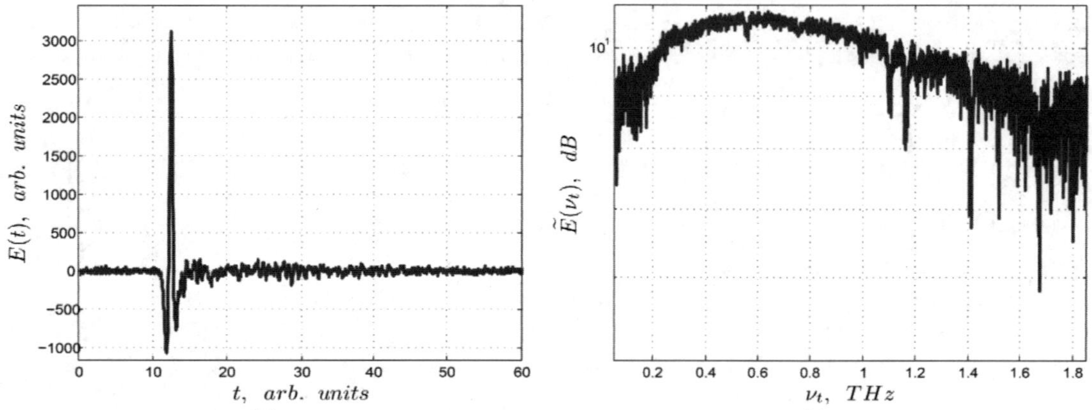

(a) **(b)**

Figure 2. Noised test TDS signal (a) and its Fourier-spectrum (b).

One could notice, that the lines of water vapor absorption located between the frequencies 1.6 and 1.8 THz (figure 1), cannot be detected based on the noised signal spectrum (figure 2). We use several criteria which help us to evaluate the quality of signal reconstruction and to compare the results of de-noising with different wavelet-bases and different thresholding techniques:

- The standard deviation criterion:

$$\sigma(\psi_{a,b}(t),T) = \sqrt{\frac{1}{N}\sum_{i=1}^{i=K}\left|\tilde{s}_{init}(v_{t_i}) - \tilde{s}_{rec}(v_{t_i})\right|^2}, \tag{11}$$

where v_t is the frequency in Fourier-domain, $\tilde{s}_{init}(v_t)$ is the initial waveform Fourier-spectrum before noise addition, $\tilde{s}_{rec}(v_t)$ is the Fourier-spectrum of the reconstructed waveform.

- The line minimum location criterion:

$$\delta_{v_t}(\psi_{a,b}(t),T) = v_{ti} - v'_{ti}, \tag{12}$$

where v_{ti} is the location of the water vapor absorption line in the Fourier-spectrum of the initial signal, and v'_{ti} is the location of this line in the Fourier spectrum of the de-noised signal.

- The generalize optimal filtration criterion:

$$I(\psi_{a,b}(t),T) = \xi \cdot \sigma(\psi_{a,b}(t),T) + \eta \cdot \delta_{v_t}(\psi_{a,b}(t),T), \tag{13}$$

where ξ and η are coefficients to regulate the impact of criteria $\sigma(\psi_{a,b}(t),T)$ and $\delta_{v_t}(\psi_{a,b}(t),T)$, respectively. We consider equal coefficients here: $\xi = \eta$.

Calculation of the listed criterion has been produced for listed threshold types and for several optimal (table 1) and non-optimal wavelet bases. The results of calculations are presented in the figure 3.

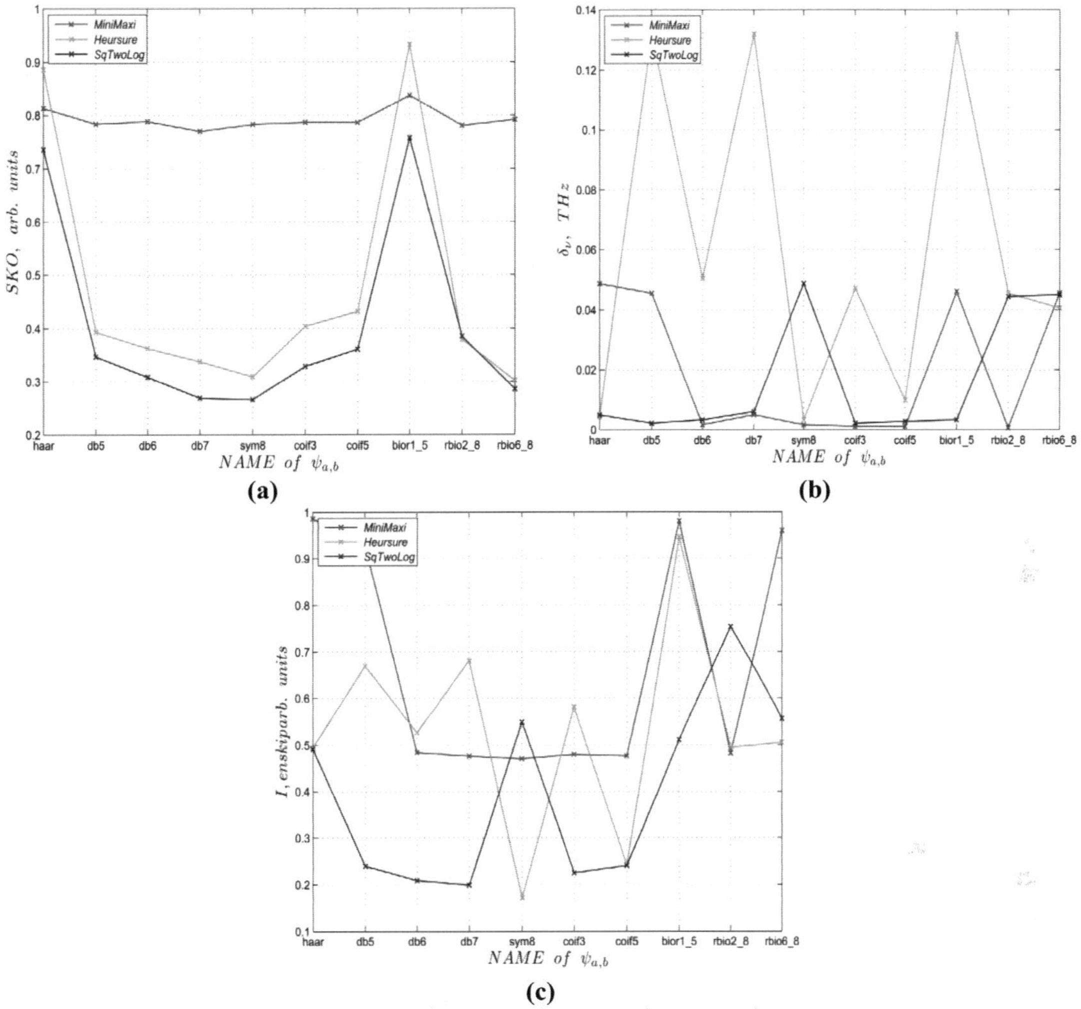

Figure 3. The values of the (a) $\sigma(\psi_{a,b}(t),T)$, (b) $\delta_{\nu_t}(\psi_{a,b}(t),T)$, and (c) $I(\psi_{a,b}(t),T)$ optimal filtration technique criteria obtained for different wavelet bases and different filtration techniques.

According to the results of calculations the best combination of the wavelet basis and wavelet-domain thresholding technique is *'sym8'* basis and *'Heursure'* threshold. Also *'db6'+'SqTwoLog'*, *'db7'+'SqTwoLog'*, *'coif3'+'SqTwoLog'*, *'coif5'+'SqTwoLog'*, and *'db6'+'Heursure'* basis and threshold combinations provide good de-noising.

The results of de-noising using the optimal combination of basis and threshold (*'sym8'* and *'Heursure'*, respectively) and the non-optimal one are presented in the figure 4.

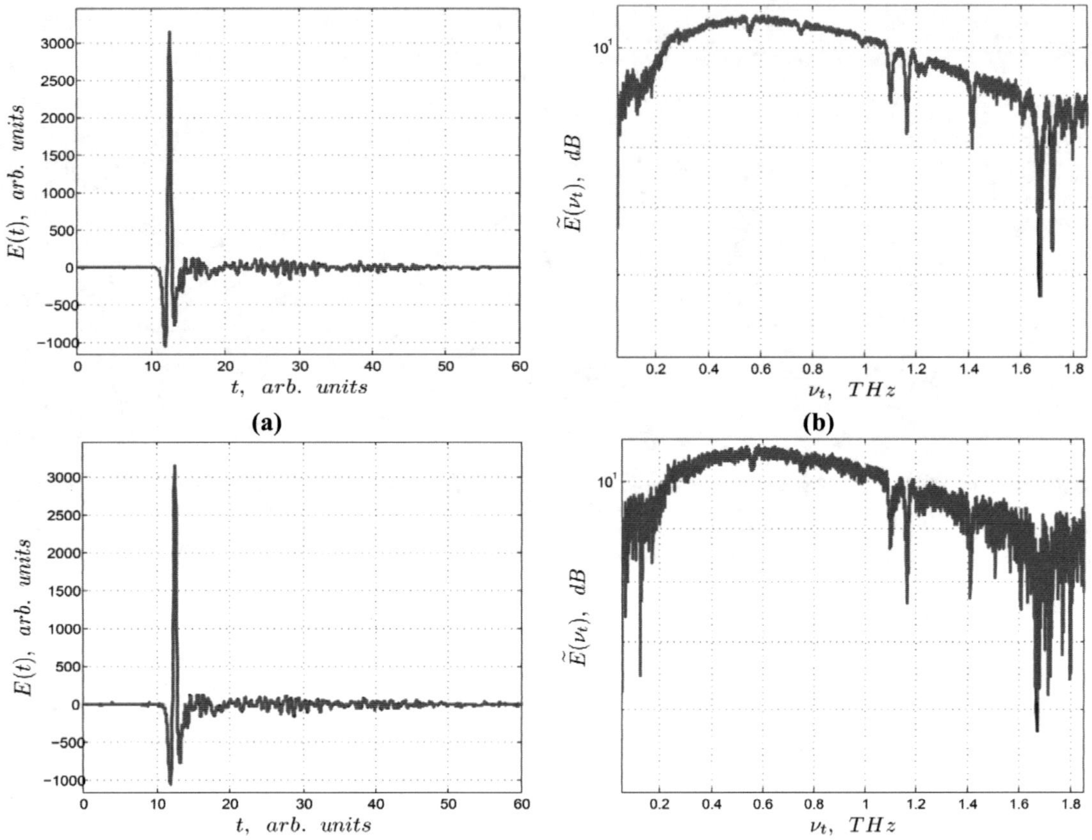

(c) (d)

Figure 4. The results of the test signal de-noising:(a,b) corresponds to the optimal filtration procedure (*'sym8'* wavelet basis and *'Heursure'* threshold), (c,d) corresponds to the non-optimal filtration (*'bior1.5'* wavelet basis and *'Heursure'* threshold)

One can note great difference between the results of optimal and non-optimal wavelet-domain de-noising implementation. The sharp peaks of the curve (figure 4(b)), corresponding to the lines of the vaporised water absorption between 1.6 and 1.8 *THz*, are clearly visible. The average topology of the reconstructed spectral curve is very close to the topology of the ideal curve presented in the figure 1. The results of the non-optimal filtration implementation contain a lot of distortions and cannot be used for correct signal analysis.

Conclusions

This paper presents the results of the wavelet-domain de-noising procedure implementation for THz pulsed spectroscopy signal processing. The problems of optimal wavelet basis and wavelet-domain filtration procedure selection are considered and solved. Novel approach for automatic construction of wavelet-domain de-noising procedure is offered.

Acknowledgments

This work was supported by grants from the Russian Foundation for Basic Research (RFBR 14-08-00175, RFBR 14-08-31124, RFBR 14-08-31767).

References

[1] Lee Y-S *Principles of Terahertz Science and Technology* (Springer, 2009) 347p

[2] Duvillaret L, Garet F, Coutaz J-L 1999 Highly precise determination of Optical Constants and sample thickness in Terahertz Time-Domain Spectroscopy *Appl. Opt.* **38** 409 – 415

[3] Dorney T, Baraniuk R, Mittleman D 2001 Material parameter estimation with Terahertz Time-Domain Spectroscopy *J. Opt. Soc. Am. A* **18** 1562 – 1571

[4] Pupeza I, Wilk R, Koch M 2007 Highly accurate optical material parameter determination with THz time-domain spectroscopy *Opt. Express* **15**(7) 4335 – 4350

[5] Zaytsev K I, Karasik V E, Fokina I N, Alekhnovich V I 2013 Invariant embedding technique for medium permittivity profile reconstruction using terahertz time-domain spectroscopy *Opt. Eng.* **52**(6) 068203-1 – 068203-10

[6] Alekhnovich V I, Zaytsev K I, Karasik V E 2013 Medium dielectric permittivity profile reconstruction using terahertz time-domain spectroscopy *Herald of Bauman Moscow State Technical University. Series: "Natural sciences"* **2** 50-66

[7] Mittleman D M, Hunsche S, Boivin L, Nuss M C 1997 T-ray tomography *Opt. Lett.* **22**(12) 904 – 906

[8] Naftaly M, Dudley R 2009 Methodologies for determining the dynamic ranges and signal-to-noise ratios of terahertz time-domain spectrometers *Opt. Lett.* **34**(8) 1213-1215

[9] Mallat S A 2009 *Wavelet Tour of Signal Processing* Third Edition: The Sparse Way. (Academic Press)

[10] Daubechies I 1992 *Ten Lectures on Wavelets*. (Society for Industrial and Applied Mathematics (SIAM))

[11] Pei J, Ye P, Xie W 2009 Optimal wavelet analysis for THz-TDS pulse signals *Proc. SPIE* **7277** 727708-1 – 727708-10

[12] Deng Y, Sun Q, Liu F, Wang Ch, Xing Q 2010 Terahertz Time-resolved Spectroscopy with Wavelet-transform *3rd International Congress on Image and Signal Processing (CISP2010)* 3462 – 3464

[13] Kim Y, Jin K H, Ye J Ch, Ahn J, Yee D-S 2011 Wavelet Power Spectrum Estimation for High-resolution Terahertz Time-domain Spectroscopy *J. Opt. Soc. Korea* **15**(1) 103-108

AUTHOR INDEX

AUTHOR INDEX